SpringerBriefs in Mathematics

SpringerBriefs in Mathematics showcases expositions in all areas of mathematics and applied mathematics. Manuscripts presenting new results or a single new result in a classical field, new field, or an emerging topic, applications, or bridges between new results and already published works, are encouraged. The series is intended for mathematicians and applied mathematicians.

SBMAC SpringerBriefs

The **SBMAC SpringerBriefs** series publishes relevant contributions in the fields of applied and computational mathematics, mathematics, scientific computing, and related areas. Featuring compact volumes of 50 to 125 pages, the series covers a range of content from professional to academic.

The Brazilian Society of Computational and Applied Mathematics (Sociedade Brasileira de Matemática Aplicada e Computacional – SBMAC) is a professional association focused on computational and industrial applied mathematics. The society is active in furthering the development of mathematics and its applications in scientific, technological, and industrial fields. The SBMAC has helped to develop the applications of mathematics in science, technology, and industry, to encourage the development and implementation of effective methods and mathematical techniques for the benefit of science and technology, and to promote the exchange of ideas and information between the diverse areas of application.

http://www.sbmac.org.br/

Carlos Hoppen • David P. Jacobs • Vilmar Trevisan

Locating Eigenvalues in Graphs

Algorithms and Applications

 Springer

Carlos Hoppen (iD)
Instituto de Matemática e Estatística
Universidade Federal do Rio Grande do Sul
Porto Alegre, Rio Grande do Sul, Brazil

David P. Jacobs (iD)
School of Computing
Clemson University
Clemson, SC, USA

Vilmar Trevisan (iD)
Instituto de Matemática e Estatística
Universidade Federal do Rio Grande do Sul
Porto Alegre, Rio Grande do Sul, Brazil

This work was supported by Conselho Nacional de Desenvolvimento Científico e Tecnológico

ISSN 2191-8198   ISSN 2191-8201   (electronic)
SpringerBriefs in Mathematics
ISBN 978-3-031-11697-1   ISBN 978-3-031-11698-8   (eBook)
https://doi.org/10.1007/978-3-031-11698-8

Mathematics Subject Classification: 15A18, 05C85, 05C05, 68W40, 68P05

This Springer imprint is published by the registered company Springer Nature Switzerland AG
The registered company address is: Gewerbestrasse 11, 6330 Cham, Switzerland

À Prof^a. Nair Abreu, cujo entusiasmo e generosidade nos trouxe à Teoria Espectral dos Grafos.

Ao Prof. Underwood Dudley, por nos revelar a beleza das demonstrações.

V. Trevisan também dedica este livro à Eliana, sua companheira de vida.

Preface

Spectral graph theory is a research area that combines tools and concepts of linear algebra and combinatorics. Its main goal is to determine properties of a graph through the eigenvalues and eigenvectors of matrices associated with it.

Perhaps surprisingly, eigenvalues and eigenvectors turn out to be intimately connected with the structure of a graph. In terms of applications, they have proved to be useful for isomorphism testing and embedding graphs in the plane, for graph partitioning and clustering, as topological descriptors for networks and molecules, in the geometric description of data sets, and in the design of efficient networks, just to mention a few. In a purely mathematical perspective, the study of graph spectra has led to a myriad of open problems, ranging from the construction of graphs with a given set of eigenvalues to extremal problems that ask for a characterization of graphs that maximize or minimize some spectral parameter.

Of course, computing these eigenvalues and eigenvectors is a necessary step in any such application. Since eigenvalues are the roots of a polynomial, in general we cannot expect to find simple expressions for these roots. However, there are numerical algorithms that allow us to approximate them with any desired precision in polynomial time.

In 2011 a linear time algorithm was discovered for deciding how many eigenvalues of the adjacency matrix of a tree lie in any interval. In addition to being fast and easy to implement, this algorithm is surprisingly simple and, at first glance, does not seem to be acting on matrices. Its underlying idea was soon extended to deal with other matrices and graph classes, giving rise to the systematic study of what we now call *eigenvalue location algorithms*.

More than being a practical way of approximating eigenvalues, these algorithms have shown to be instrumental to settle theoretical questions involving the distribution of eigenvalues in graphs in a given class. In particular, an eigenvalue location algorithm was the main tool for the recent solution of a difficult conjecture involving the distribution of Laplacian eigenvalues in trees.

In this book, we survey the evolution of eigenvalue location algorithms in an organized and unified way, starting with algorithms for trees and other well-known graph classes, such as cographs, and showing how they motivated more recent algorithms that may be applied to arbitrary graphs, but whose efficiency depends on the existence of a graph decomposition of low complexity. While they are vastly deeper than the simple tree algorithm, we wish to convince the readers that they are similar in spirit.

This book is intended for graduate students and researchers in spectral graph theory. We have strived to make it as self-contained as possible. This has led to a book that combines a compact introduction to spectral graph theory with a discussion of structural decompositions such as tree decompositions and clique decompositions, which is rarely explored by books in this area. We hope that it may be used as a concise introduction to eigenvalue location algorithms for researchers who wish to know more about eigenvalues associated with graphs.

Porto Alegre, Brazil   Carlos Hoppen
Clemson, SC, USA   David P. Jacobs
Porto Alegre, Brazil   Vilmar Trevisan
April 2022

Acknowledgments

The authors are grateful to the Brazilian funding agencies CAPES (Coordenação de Aperfeiçoamento de Pessoal de Nível Superior), CNPq (Conselho Nacional de Desenvolvimento Científico e Tecnológico), and FAPERGS (Fundação de Amparo à Pesquisa do Estado do Rio Grande do Sul). For many years, these agencies have fostered the collaboration between Brazilian and international scientists from all over the world. In particular, they funded several research visits leading to results described in this book.

Our thanks to SBMAC (Sociedade Brasileira de Matemática Aplicada e Computacional) and to Springer Nature for sponsoring this book. A special thank you to Celina de Figueiredo and Paulo Silva, the editors of SBMAC Springer Briefs, and to Robinson dos Santos, from the São Paulo office of Springer Nature.

We are also thankful to three anonymous reviewers, whose careful reading and thoughtful considerations have led to a better book.

Many of the figures in this book were made using TikZiT, a program for drawing graphs, making the construction of such figures simple and straightforward.

Contents

Chapter 1
Introduction

Eigenvalues are numbers that are associated with matrices and have been of interest to mathematicians for over two hundred years. They seem to come up in almost every branch of science, including biology [1], chemistry [21], epidemiology [14, 22], geology [20], data science [13], dynamical systems [7], and network design [3]. As of this writing, a search for the word *eigenvalue* in the MathSciNet database of the American Mathematical Society returns 87,293 matches.

Spectral graph theory seeks to study graphs through the eigenvalues and eigenvectors of matrices associated with them. For example, just looking at the multiset of eigenvalues, or *spectrum*, of the adjacency and Laplacian matrices associated with a graph, we are able to tell its number of vertices, edges, and triangles, how many components it has, whether it is bipartite or not, among many other things. Even parameters that are hard to compute, such as the domination number, the chromatic number or the independence number of a graph G, can often be bounded in terms of certain eigenvalues of matrices associated with G.

The eigenvalues of a matrix are precisely the roots of the so-called *characteristic polynomial*, which, for an $n \times n$ matrix, has degree n and can generally be constructed efficiently. The problem of finding the roots of a polynomial, also known as solving an algebraic equation, is very old and has played a fundamental role in the evolution of mathematics since ancient history. The quest for algorithms that find the solutions to these equations continued through the middle ages and the first centuries of modern history and culminated with the work of 19th century mathematicians such as Abel and Galois, who proved groundbreaking results showing that there is no general formula for the roots of polynomials with integer coefficients having degree five or more. Nowadays, there are powerful numerical algorithms that are able to approximate the eigenvalues with any given precision in cubic time [16].

In this book, we address the problem of computing eigenvalues in terms of *eigenvalue location*, by which we mean determining the number of eigenvalues of

C. Hoppen et al., *Locating Eigenvalues in Graphs*, SpringerBriefs in Mathematics, https://doi.org/10.1007/978-3-031-11698-8_1

a symmetric matrix[1] that lie in any given real interval. For example, given a tree T with n vertices, suppose we could infer that there are five eigenvalues of the adjacency matrix of T that are greater than 2, by simply counting the number of positive values in a multiset of n numbers. Now suppose we could also calculate that there are four eigenvalues greater than 3. Then the tree must have exactly one eigenvalue λ in the interval $(2, 3]$. As we shall see, this is actually possible using one of our eigenvalue location algorithms for trees, and calculations can be performed on the tree itself. In theory, we could further approximate λ by divide and conquer. That is, we could subdivide the interval and determine if it resides in $(2, \frac{5}{2}]$ or $(\frac{5}{2}, 3]$, and so forth, until we approximate λ with the desired precision.

As it turns out, our eigenvalue location algorithms are based on the notion of *matrix congruence* and on a classical linear algebra result known as Sylvester's Law of Inertia. Moreover, either they are designed for graphs in a particular class and exploit some special features of this class, or they rely on a structural decomposition of the input graph. In both cases, a crucial feature is that the algorithms are very fast, running in linear time for all graphs in the said class or for all graphs such that there is a structural decomposition of low complexity, in a way that will be defined precisely. More generally, we should mention that matrix congruence is a basic relation in linear algebra and that our algorithms operate by finding a diagonal matrix D that is congruent to an input matrix M. Some of them, particularly in Chaps. 3 and 6, may be applied to large classes of symmetric matrices and may be used in applications of congruence beyond eigenvalue location.

However, the benefit of eigenvalue location algorithms goes far beyond finding eigenvalues of a *particular* graph and often allows us to derive properties of an entire class, making these algorithms valuable theoretical tools. Here are two examples that will be explained more in detail later in the book. First, by analyzing the algorithm for the famous class of graphs known as cographs, one can show that every cograph has the same number of eigenvalues in the intervals $(-1, \infty)$ and $[0, \infty)$, thus concluding that cographs have no eigenvalues in $(-1, 0)$. Second, one can show through the analysis of the tree algorithm that at least half of its Laplacian eigenvalues are in the interval $[0, 2)$. In fact, a well-known open problem in graph theory stated that, in any tree of order n, at least half of its Laplacian eigenvalues lie in the interval $[0, 2 - \frac{2}{n})$. The validity of this statement has been recently proved using the same general principle used for $[0, 2)$, even though the technical details are more involved and are beyond the scope of this book.

Our book is structured to give the reader the necessary mathematical background to understand the eigenvalue location algorithms, their underlying graph representations, and decompositions and to provide interesting applications. The systematic study of these algorithms is now over a decade old, allowing us to highlight commonalities and general principles that were absent from the early papers. We also aimed at a book that was as self-contained as possible, which gives it the

---

[1] We note that most classical matrices associated with graphs are symmetric, and therefore their eigenvalues are always real.

distinctive feature of providing a compact introduction both to spectral graph theory and to structural decompositions that are not always explored by the community that works in this area. Of course, there are many excellent books that focus on spectral graph theory and treat the subject under a deeper and more encyclopedic point of view, such as [6, 8–10, 15, 19]. There is also a host of great books that delve deep into graph decompositions, and many of them will be mentioned in the next chapters.

In Chap. 2, we provide definitions and terminology for the mathematical theory upon which the algorithms are based, including graph theory and linear algebra. Readers with experience in these areas might want to skip it and come back if some clarification is needed. Chapter 3 describes the linear time eigenvalue location algorithm for trees. In addition to its historical significance for being the first algorithm of this type, it is surprisingly simple and elegant. When running the algorithm, one does not see immediately that it acts on a matrix. Moreover, calculations may be easily performed by hand. This algorithm will serve as a template for algorithms in the next chapters and will allow us to state an eigenvalue location principle.

In Chap. 4, we move into structural graph theory and discuss graph classes and graph decompositions that will be used in subsequent chapters. In particular, we present the class of cographs and its many characterizations. We also discuss tree decompositions and clique decompositions of graphs and specific versions of these decompositions that are used in our algorithms, known as nice tree decompositions and slick clique expressions. Chapters 5, 6 and 7 deal with the eigenvalue location algorithm for cographs and the algorithms based on a tree decomposition and a clique decomposition, respectively. The first algorithm runs in linear time, while the other two run in time $O(k^2 n)$, where k denotes the *width* of the decomposition, and are therefore linear time algorithms for graphs with bounded width, provided that the decomposition is given as part of the input. These three chapters are essentially independent, but it is worth mentioning that the class of cographs is precisely the class of graphs with slick clique-width equal to 1, and the algorithm of Chap. 7 may therefore be viewed as an extension of the algorithm of Chap. 5.

Chapter 8 contains material that is not available in other sources and considers an application of the algorithm of Chap. 7 to the class of distance-hereditary graphs, a well-known graph class whose elements have slick clique-width at most 2. To do this, we relate the algorithm with a structural representation of these graphs using the so-called *pruning trees*. Chapter 8 also discusses properties of graphs with bounded slick clique-width more broadly.

To conclude the introduction, we would like to mention that this book does not present all eigenvalue location algorithms that may be found in the literature. There are algorithms that deal specifically with threshold graphs [18], unicyclic graphs [5], chain graphs [2], and generalized lollipop graphs [11]. Similar algorithms have also been designed for classes of signed graphs [4]. Moreover, the book does not address eigenvalue location algorithms that take advantage of the structure of distance matrices [12, 17].

References

1. Akemann, G., Baake, M., Chakarov, N., Krüger, O., Mielke, A., Ottensmann, M., Werde-hausen, R.: Territorial behaviour of buzzards versus random matrix spacing distributions. J. Theoret. Biol. **509**, 110475, 7 (2021). https://doi.org/10.1016/j.jtbi.2020.110475
2. Alazemi, A., Anđelić, M., Simić, S.K.: Eigenvalue location for chain graphs. Linear Algebra Appl. **505**, 194–210 (2016). https://doi.org/10.1016/j.laa.2016.04.030
3. Becker, C.O., Pequito, S., Pappas, G.J., Preciado, V.M.: Network design for controllability metrics. IEEE Trans. Control Netw. Syst. **7**(3), 1404–1415 (2020)
4. Belardo, F., Brunetti, M., Trevisan, V.: Locating eigenvalues of unbalanced unicyclic signed graphs. Appl. Math. Comput. **400**, 126082 (2021)
5. Braga, R.O., Rodrigues, V.M., Trevisan, V.: Locating eigenvalues of unicyclic graphs. Appl. Anal. Discrete Math. **11**, 273–298 (2017)
6. Brouwer, A.E., Haemers, W.H.: Spectra of Graphs. Universitext. Springer, New York (2012). https://doi.org/10.1007/978-1-4614-1939-6
7. Chu, K.w.E., Govaerts, W., Spence, A.: Matrices with rank deficiency two in eigenvalue problems and dynamical systems. SIAM J. Numer. Anal. **31**(2), 524–539 (1994). https://doi.org/10.1137/0731028
8. Chung, F.R.K.: Spectral graph theory, *CBMS Regional Conference Series in Mathematics*, vol. 92. Published for the Conference Board of the Mathematical Sciences, Washington, DC; by the American Mathematical Society, Providence, RI (1997)
9. Cvetković, D., Rowlinson, P., Simić, S.: Eigenspaces of graphs. Encyclopedia of Mathematics and Its Applications, vol. 66. Cambridge University Press, Cambridge (1997). https://doi.org/10.1017/CBO9781139086547
10. Cvetković, D., Rowlinson, P., Simić, S.: An introduction to the theory of graph spectra. London Mathematical Society Student Texts, vol. 75. Cambridge University Press, Cambridge (2010)
11. Del-Vecchio, R.R., Jacobs, D.P., Trevisan, V., Vinagre, C.T.M.: Diagonalization of generalized lollipop graphs. Elec. Notes Discrete Math. **50**, 41–46 (2015)
12. Du, Z.: A diagonalization algorithm for the distance matrix of cographs. IEEE Access **6**, 74931–74939 (2018)
13. Feng, S., Cetinkaya, A., Ishii, H., Tesi, P., De Persis, C.: Networked control under DoS attacks: tradeoffs between resilience and data rate. IEEE Trans. Automat. Control **66**(1), 460–467 (2021)
14. Frank, T.D.: COVID-19 interventions in some European countries induced bifurcations stabilizing low death states against high death states: an eigenvalue analysis based on the order parameter concept of synergetics. Chaos Solitons Fractals **140**, 110194, 7 (2020). https://doi.org/10.1016/j.chaos.2020.110194
15. Godsil, C., Royle, G.: Algebraic graph theory. Graduate Texts in Mathematics, vol. 207. Springer, New York (2001). https://doi.org/10.1007/978-1-4613-0163-9
16. Golub, G.H., van der Vorst, H.A.: Eigenvalue computation in the 20th century. Numerical analysis 2000, Vol. III. Linear algebra, pp. 35–65 (2000). https://doi.org/10.1016/S0377-0427(00)00413-1
17. Jacobs, D.P., Trevisan, V., Tura, F.: Distance eigenvalue location in threshold graphs. Proc. of DGA, pp. 1–4 (2013a)
18. Jacobs, D.P., Trevisan, V., Tura, F.: Eigenvalue location in threshold graphs. Linear Algebra Appl. **439**(10), 2762–2773 (2013b). https://doi.org/10.1016/j.laa.2013.07.030
19. Li, X., Shi, Y., Gutman, I.: Graph Energy. Springer, New York (2012). https://doi.org/10.1007/978-1-4614-4220-2
20. Schoof, C.: Cavitation on deformable glacier beds. SIAM J. Appl. Math. **67**(6), 1633–1653 (2007). https://doi.org/10.1137/050646470
21. Sperb, R.P.: On an eigenvalue problem arising in chemistry. Z. Angew. Math. Phys. **32**(4), 450–463 (1981). https://doi.org/10.1007/BF00955622
22. Wang, J., Chen, Y.: Threshold dynamics of a vector-borne disease model with spatial structure and vector-bias. Appl. Math. Lett. **100**, 106052, 7 (2020). https://doi.org/10.1016/j.aml.2019.106052

Chapter 2
Preliminaries

2.1 Graph Theory Review

In this section, we review some graph theoretical ideas needed to understand the algorithms presented later and refer to [2, 3, 8, 11, 17, 24] for a more exhaustive treatment of the subject. This monograph concerns itself with *graphs*, or sometimes called *simple graphs*. Formally, a graph G is an ordered pair (V, E), where V is a finite set of *vertices or nodes* and E is a set of two-element subsets of V called *edges*. The *order* of the graph is $n = |V|$. The *size* of G is $m = |E|$. It is easier to visualize a graph as circles for the vertices and line segments for the edges. Figure 2.1 depicts a graph of order five.

If $e = \{v, w\} \in E$, we say that vertices v and w are *adjacent* and that edge e is *incident* with these vertices. For a vertex v, the set of vertices that are adjacent to v is the *neighborhood* (or *open neighborhood*) $N(v)$ of v. The members of $N(v)$ are said to be *neighbors* of v. The *closed neighborhood* of v is defined as $N[v] = N(v) \cup \{v\}$. Since each edge has cardinality two, a vertex is never adjacent to itself. Since E is a set, there is at most one edge between any pair of vertices. In Fig. 2.1, $N(a) = \{b, c\}$ and $N[a] = \{a, b, c\}$.

If $G = (V, E)$ and $H = (W, F)$ are graphs such that $W \subseteq V$ and $F \subseteq E$, then H is called a *subgraph* of G. If G is the graph in Fig. 2.1, let $e_1 = \{a, b\}$, $e_2 = \{c, d\}$, and $F = \{e_1, e_2\}$. Then $H = (V, F)$ is a subgraph of G. Now let $S \subseteq V$. The *induced subgraph* $G[S]$ is the graph (S, F), where $F \subseteq E$ is the set of edges $\{u, v\}$ such that $u, v \in S$. If G is the graph in Fig. 2.1 and $S = \{b, c, e\}$, then the induced subgraph $G[S]$ has only one edge $\{c, e\}$. If H is an induced subgraph of G, we write $H \lhd G$. The subgraph H above is not an induced subgraph. However, the subgraph with vertex set $\{c, d, e\}$ whose edges are $\{c, e\}$, $\{c, d\}$, and $\{d, e\}$ is an induced subgraph.

Given a graph $G = (V, E)$ and a vertex $v \in V$, we let $G - v$ be the graph obtained from G by deleting vertex v (and the edges incident with it). In other

© The Author(s), under exclusive license to Springer Nature Switzerland AG 2022
C. Hoppen et al., *Locating Eigenvalues in Graphs*, SpringerBriefs in Mathematics,
https://doi.org/10.1007/978-3-031-11698-8_2

Fig. 2.1 A graph with order
5 containing cycles

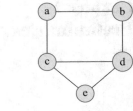

Fig. 2.2 A tree of order 9
with six leaves

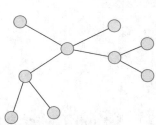

words, $G - v = G[V \setminus \{v\}]$. More generally, if $S \subset V$, we write $G - S = G[V \setminus S]$
for the induced subgraph of G obtained by deleting vertices in S.

Another well-known graph operation is *edge contraction*. Given a graph $G = (V, E)$ and an edge $e = \{x, y\} \in E$, let G/e be the graph obtained from G by
contracting e, that is, the graph obtained from G by deleting x and y and then adding
a new vertex v_e adjacent to all neighbors of x or y in G (except x and y themselves).

A *path* in a graph $G = (V, E)$ from w_1 to w_k is a sequence of distinct vertices

$$w_1, w_2, \ldots, w_k, \tag{2.1}$$

where for each $j \in \{1, \ldots, k - 1\}$, w_j and w_{j+1} are adjacent. The graph is said
to be *connected* if there is a path between any two vertices. The graph in Fig. 2.1
is connected. A graph's *components* are its maximal connected subgraphs. In the
path (2.1) if w_1 and w_k are adjacent and $k \geq 3$, then we call

$$w_1, w_2, \ldots, w_k, w_1$$

a *cycle*. The *length* of a path or cycle is the number of edges in it. The graph in
Fig. 2.1 has cycles of length 3, 4, and 5. The *distance* $d_G(u, v)$ between vertices u
and v in a graph G is the length of a shortest path between them. The *diameter* of a
connected graph, written $diam(G)$, is the largest $d_G(u, v)$ over all pairs of vertices.

A *forest* is a graph having no cycles. A *tree* is a connected forest. It can be shown
that if $T = (V, E)$ is a tree, then $|E| = |V| - 1$. Figure 2.2 depicts a tree of order
nine and diameter four.

The *degree* of a vertex v, denoted $\deg(v)$, is the number of neighbors it has, that
is, $|N(v)|$. Clearly, $0 \leq \deg(v) \leq n - 1$ for a graph of order n. If $\deg(v) = 0$, v is
called an *isolate*. If $\deg(v) = n - 1$, then v is adjacent to every other vertex in the

graph. A vertex of degree one in a tree is called a *leaf* so the tree in Fig. 2.2 has six leaves. As it turns out, every tree with $n \geq 2$ vertices contains two or more leaves.

Two graphs $G = (V, E)$ and $H = (W, F)$ are said to be *isomorphic* if there is an edge-preserving bijection $\phi \colon V \to W$, namely a bijection such that $\{\phi(u), \phi(v)\} \in F$ if and only if $\{u, v\} \in E$. The function ϕ is called an *isomorphism* between G and H. Consider the relation $\sim$ defined by $G \sim H$ if G and H are isomorphic. It is easy to see that this is an *equivalence relation*, meaning it is symmetric, reflexive, and transitive. In this relation, two graphs lie in the same equivalence class if and only if they are equal up to vertex names (or labels). The equivalence classes are called *unlabeled graphs* and may be depicted as graphs whose vertices have no distinct identification except for their connectivity. Figure 2.1 depicts a labeled graph, while Fig. 2.2 depicts an unlabeled graph.

Unlabeled graphs have intuitive appeal, as they preserve the abstract structure of a graph, while ignoring labels and particular representations, such as embeddings in a space or on a surface. Some special graphs, of order n, are the *path* P_n, a tree with two leaves, the *star* S_n, a tree with $n - 1$ leaves, the *cycle* C_n, a connected graph in which every vertex has degree two, and the *complete graph* K_n in which every vertex is adjacent to every other vertex. These are illustrated in Fig. 2.3. Note that these graphs are unlabeled, as they have been defined only in terms of their edge structure. There are $\binom{n}{2} = \frac{n(n-1)}{2}$ two-element sets of V. The *complement* $\overline{G}$ of a graph $G = (V, E)$ is the graph $(V, \overline{E})$ containing exactly the edges $e \notin E$. If $|E| = m$, then $|\overline{E}| = \binom{n}{2} - m$. Note that $\overline{K_n}$ has no edges.

Unlabeled graphs are intertwined with the concept of a *graph property* or *graph invariant*, which is a property of graphs that is preserved under any isomorphism. Connectedness is such a graph invariant.

Another well-known graph invariant is the *chromatic number* $\chi(G)$ of a graph $G = (V, E)$, which is the minimum number k of colors such that a color may be assigned to each vertex of G so that any adjacent vertices have different colors. Formally, it is the minimum integer k such that there exists a function $f \colon V \to \{1, \ldots, k\}$ such that $f(u) \neq f(v)$ for all $\{u, v\} \in E$.

In a graph $G = (V, E)$, a vertex set S is called a *dominating set* if every $v \in V - S$ is adjacent to some member of S. The *domination number* $\gamma(G)$ of G is the minimum cardinality of a dominating set.

Two other graph invariants that will be mentioned in this book are the *independence number* $\alpha(G)$ and the *clique number* $\omega(G)$ of a graph G. The former is the maximum size of a set $S \subseteq V$ such that $G[S]$ has no edges. The latter is the

Fig. 2.3 The graphs P_5, S_5, C_6, and K_4

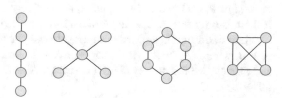

maximum size of a set $S \subseteq V$ such that $G[S]$ is a *clique*, meaning that there are edges between all pairs of vertices. One can show that $\alpha(\overline{G}) = \omega(G)$.

A graph $G = (V, E)$ is *bipartite* if $V = U \cup W$, $U \cap W = \emptyset$, and each edge is incident with a vertex in U and a vertex in W. It turns out that a graph is bipartite if and only if it does not contain any cycle of odd length as a subgraph, so that trees are bipartite. Among graphs with edges, bipartite graphs are exactly those with $\chi(G) = 2$.

2.2   Linear Algebra Review

Here we give an overview of linear algebra and refer to [18] for a deeper treatment of the subject. A *vector space* V over a field F is an algebraic structure having two operations:

(a) A binary operator $+$ such that $(V, +)$ forms an Abelian group.
(b) A scalar multiplication $F \times V \to V$ such that for all $\alpha, \beta \in F$ and all $v, w \in V$ one has $\alpha(v + w) = \alpha v + \alpha w$, $(\alpha + \beta)v = \alpha v + \beta v$, $(\alpha\beta)v = \alpha(\beta v)$, and $1v = v$. We suppress the scalar multiplication operator.

For our purposes, the field will be the real numbers $\mathbb{R}$. Our vector spaces will be the set of length n column vectors with real entries, denoted $\mathbb{R}^{n \times 1}$, and the set of length n row vectors with real entries, denoted $\mathbb{R}^{1 \times n}$. We obtain a *subspace* W by taking a set $S \subseteq V$ and forming all finite sums and scalar products, denoted $W = \langle S \rangle$. The *dimension* of W, written $\dim(W)$, is the minimum cardinality of a set S that *spans* W, that is, such that any element of W may be expressed as $\sum_{i=1}^{k} a_i v_i$, where $a_i \in \mathbb{R}$ and $v_i \in S$. One can show that $\dim(\mathbb{R}^{n \times 1}) = \dim(\mathbb{R}^{1 \times n}) = n$.

If U and W are subspaces of a vector space, then $U + W$ is the subspace $\{\alpha u + \beta w \mid u \in U, w \in W, \alpha, \beta \in \mathbb{R}\}$. Then

$$\dim(U + W) = \dim(U) + \dim(W) - \dim(U \cap W). \tag{2.2}$$

An $n \times m$ *matrix* M has n rows and m columns with real entries. The entry in row i and column j is denoted M_{ij}. The *transpose* M^T is the $m \times n$ matrix formed by setting $M_{ij}^T = M_{ji}$ for all i, j. In an $n \times m$ matrix, columns are considered vectors in $\mathbb{R}^{n \times 1}$, and rows are vectors in $\mathbb{R}^{1 \times m}$. We may also regard column vectors as $n \times 1$ matrices and row vectors as $1 \times m$ matrices. If $\mathbf{x} \in \mathbb{R}^{n \times 1}$, then $\mathbf{x}^T \in \mathbb{R}^{1 \times n}$ and vice versa. Let $R_i(M)$ and $C_i(M)$ denote, respectively, row i and column i of matrix M. If there is only one matrix under discussion, we write R_i or C_i.

Matrix addition is defined component-wise for two $m \times n$ matrices, that is, $C = A + B$, where $C_{ij} = A_{ij} + B_{ij}$ for each i, j. Matrix addition is clearly commutative and associative for $m \times n$ matrices, that is, $A + B = B + A$ and $A + (B + C) = (A + B) + C$.

If $\mathbf{x} \in \mathbb{R}^{1 \times n}$ and $\mathbf{y} \in \mathbb{R}^{n \times 1}$, their *dot product* (or *inner product*) is

$$\mathbf{x} \cdot \mathbf{y} = \sum_{i=1}^{n} x_i y_i.$$

Matrix multiplication is defined as follows. Let A be an $m \times k$ matrix, and let B be a $k \times n$ matrix. The product AB is an $m \times n$ matrix P, where $P_{ij} = R_i(A) \cdot C_j(B)$. If M is $n \times n$, and $\mathbf{x} \in \mathbb{R}^{n \times 1}$, then $M\mathbf{x}$ is another column vector.

An $n \times n$ matrix is called a *square* matrix. If A, B, C are $n \times n$ matrices, matrix multiplication is associative, that is, $A(BC) = (AB)C$. Note that the associative law applies to two binary operations, but it can be extended to arbitrary products. However, matrix multiplication is not commutative. That is, $AB \neq BA$ in general.

We let I_n denote the $n \times n$ *identity* matrix:

$$\begin{pmatrix} 1 & 0 & \cdots & 0 \\ 0 & 1 & \cdots & 0 \\ \vdots & & \ddots & \vdots \\ 0 & 0 & \cdots & 1 \end{pmatrix}.$$

If A is any $n \times n$ matrix, then $A = I_n A = A I_n$. The matrix A is *nonsingular* if there exists a matrix B such that $BA = AB = I_n$. In this case, we write A^{-1} for B, and A^{-1} is called the *inverse* of A. The inverse, if it exists, is unique. There is a parameter associated with square matrices, the *determinant* such that A is nonsingular if and only if $\det(A) \neq 0$. If A does not have an inverse, we say it is *singular*. If A and B are both nonsingular, then it is easy to see that their product is also nonsingular and $(AB)^{-1} = B^{-1}A^{-1}$. By induction, this can be extended to finite products as follows:

$$(A_1 A_2 \ldots A_k)^{-1} = A_k^{-1} \ldots A_2^{-1} A_1^{-1}. \tag{2.3}$$

A square matrix M is *symmetric* if $M_{ij} = M_{ji}$ for all i, j. Note the symmetry across the main diagonal. Of course M is symmetric if and only if $M = M^T$. We claim that for any square matrices A and B

$$(AB)^T = B^T A^T. \tag{2.4}$$

To see this, we observe that both sides have the same entries. The ij entry on the left side of (2.4) is $(AB)_{ij}^T = (AB)_{ji} = R_j(A) \cdot C_i(B)$. The ij entry on the right of (2.4) is $(B^T A^T)_{ij} = R_i(B^T) \cdot C_j(A^T) = (C_i(B))^T \cdot (R_j(A))^T$, and these are equal by commutativity of multiplication. By induction, this can be extended to finite products

$$(A_1 A_2 \ldots A_k)^T = A_k^T \ldots A_2^T A_1^T. \tag{2.5}$$

If M is nonsingular, then we claim that

$$(M^{-1})^T = (M^T)^{-1}. \tag{2.6}$$

To see this, using (2.4), we have $(M^{-1})^T (M^T) = (MM^{-1})^T = I_n{}^T = I_n$.
Therefore, the inverse of M^T is the left side of (2.6).

Two square matrices A and B are *similar* if there exists a nonsingular matrix C such that $A = C^{-1}BC$. If A and B are similar, we write $A \simeq B$. Similarity is an equivalence relation on the $n \times n$ matrices. This can be verified with the matrix properties above.

Two real symmetric matrices A and B are *congruent* if there exists a nonsingular matrix C such that $A = C^T BC$. If A and B are congruent, we write $A \cong B$. Congruence is also an equivalence relation on the symmetric $n \times n$ matrices. For symmetric matrices, similar matrices are congruent, but the converse is not true. We will see in Sect. 2.6 that congruence has only a finite number of equivalence classes. The matrix D is *diagonal* if $D_{ij} = 0$ for $i \neq j$. We write $\text{diag}(d_1, d_2, \ldots, d_n)$ to mean the diagonal matrix D where $D_{ii} = d_i$. In some texts, a square matrix is said to be *diagonalizable* if it is similar to a diagonal matrix. We use *diagonalization* to mean the process of finding a diagonal matrix that is *congruent* to a symmetric matrix.

Congruence and diagonalization are important ideas in this book.

We now introduce yet a third equivalence relation on matrices. Two matrices A and B are *equivalent* if there exist nonsingular matrices P and Q such that $A = PBQ$. The *rank* of a matrix is the dimension of the subspace spanned by its row vectors. It is well known that matrices are equivalent if and only if they have the same rank. One can see that similarity and congruence are refinements of equivalence.

2.3 Eigenvalues and Eigenvectors

Let M be an $n \times n$ matrix over the reals. We say a complex number λ is an *eigenvalue* for M if there exists a nonzero column vector $\mathbf{x}$ such that $M\mathbf{x} = \lambda\mathbf{x}$. The vector is called an *eigenvector* for λ. The eigenvalues are precisely the roots of the *characteristic polynomial* $\det(M - xI)$, and the *multiplicity* of an eigenvalue is the number of times it occurs as a root.[1] The *spectrum* of M is the multiset of

---

[1] Eigenvalues and eigenvectors are sometimes called *characteristic values* and *characteristic vectors*, respectively.

eigenvalues, where the number of occurrences of each eigenvalue is given by its multiplicity. It is well known that real symmetric matrices have real eigenvalues, and the eigenvectors associated with distinct eigenvalues are *orthogonal*, meaning that the dot product of two eigenvectors is zero.

When M is viewed as a linear transformation on $\mathbb{R}^{n \times 1}$ and $\lambda \neq 0$, an eigenvector **x** is mapped in the same direction by a factor of λ. Eigenvalues are difficult to calculate. Alluding to the work of E. Galois, in [20], Meyer states:

> As we have seen, computing eigenvalues boils down to solving a polynomial equation. But determining solutions to polynomial equations can be a formidable task. It was proven in the nineteenth century that it's impossible to express the roots of a general polynomial of degree five or higher using radicals of the coefficients. This means that there does not exist a generalized version of the quadratic formula for polynomials of degree greater than four, and general polynomial equations cannot be solved by a finite number of arithmetic operations involving $+, -, \times, \div, \sqrt{n}$. Unlike solving $Ax = b$, the eigenvalue problem generally requires an infinite algorithm, so all practical eigenvalue computations are accomplished by iterative methods . . .

It is well known that $\det(M) = \prod_{\lambda \in S} \lambda$, where S is the spectrum of M. One can show that a matrix is singular if and only if $0 \in S$ and if and only if $\det(M) = 0$.

Theorem 2.1 *If two matrices are similar, they must have the same eigenvalues.*

Proof Suppose $A = C^{-1} B C$ for some nonsingular C. The characteristic polynomial of A is

$$\det(A - x I_n) = \det(C^{-1} B C - x I_n)$$
$$= \det(C^{-1} B C - x C^{-1} I_n C)$$
$$= \det(C^{-1} (B - x I_n) C)$$
$$= \det(C^{-1}) \det(B - x I_n) \det(C)$$
$$= \det(B - x I_n)$$

using the fact that $\det(MN) = \det(M)\det(N)$ for any square matrices M and N and $\det(I_n) = 1$. Since A and B have the same characteristic polynomial, they must have same spectrum. □

It is easy to see that if D is a diagonal matrix, then the eigenvalues are the diagonal values. So if A is similar to D, where D is a diagonal matrix, by Theorem 2.1, we know that the eigenvalues of A are the diagonal elements of D. It is natural to ask whether the congruence of symmetric matrices may also be determined by the spectrum. This is answered in Sect. 2.6.

2.4 Elementary Matrices and Operations

Elementary row operations are used in many algorithms such as reducing a matrix to a desired form. We can also apply elementary operations on columns.

A type I elementary row (column) operation on a matrix involves interchanging two rows (columns). The left side of (2.7) shows a type I row operation that interchanges the first two rows, denoted as $R_1 \leftrightarrow R_2$. The right side shows interchanging the first two columns, denoted as $C_1 \leftrightarrow C_2$.

$$\begin{pmatrix} 1 & 2 & 3 \\ 4 & 5 & 6 \\ 7 & 8 & 9 \end{pmatrix} \rightarrow \begin{pmatrix} 4 & 5 & 6 \\ 1 & 2 & 3 \\ 7 & 8 & 9 \end{pmatrix} \qquad \begin{pmatrix} 1 & 2 & 3 \\ 4 & 5 & 6 \\ 7 & 8 & 9 \end{pmatrix} \rightarrow \begin{pmatrix} 2 & 1 & 3 \\ 5 & 4 & 6 \\ 8 & 7 & 9 \end{pmatrix}. \qquad (2.7)$$

An *elementary matrix* is obtained by performing the *same* elementary operation on the identity matrix I_n. Below are type I elementary matrices corresponding to the operations above. While there is a big difference between the two operations in (2.7), there is really no difference between their corresponding elementary matrices.

$$E_{R_1 \leftrightarrow R_2} = E_{C_1 \leftrightarrow C_2} = \begin{pmatrix} 0 & 1 & 0 \\ 1 & 0 & 0 \\ 0 & 0 & 1 \end{pmatrix}.$$

A type II elementary row (column) operation involves multiplying a row (column) by any nonzero scalar. The left side of (2.8) shows a type II row operation in which the third row has been multiplied by $\frac{1}{3}$, denoted as $R_3 \leftarrow \frac{1}{3}R_3$. The right side of (2.8) shows a type II column operation where the third column has been multiplied by $\frac{1}{3}$, denoted by $C_3 \leftarrow \frac{1}{3}C_3$.

$$\begin{pmatrix} 1 & 2 & 3 \\ 4 & 5 & 6 \\ 7 & 8 & 9 \end{pmatrix} \rightarrow \begin{pmatrix} 1 & 2 & 3 \\ 4 & 5 & 6 \\ \frac{7}{3} & \frac{8}{3} & 3 \end{pmatrix} \qquad \begin{pmatrix} 1 & 2 & 3 \\ 4 & 5 & 6 \\ 7 & 8 & 9 \end{pmatrix} \rightarrow \begin{pmatrix} 1 & 2 & 1 \\ 4 & 5 & 2 \\ 7 & 8 & 3 \end{pmatrix}. \qquad (2.8)$$

Performing these same operations on I_3, we obtain the following type II elementary matrices:

$$E_{R_3 \leftarrow \frac{1}{3}R_3} = E_{C_3 \leftarrow \frac{1}{3}C_3} = \begin{pmatrix} 1 & 0 & 0 \\ 0 & 1 & 0 \\ 0 & 0 & \frac{1}{3} \end{pmatrix}.$$

A type III elementary row (column) operation involves adding a multiple of a row (column) to another row (column). Below are two type III operations. In the first operation, $\frac{1}{2}R_2$ is added to R_1 written as $R_1 \leftarrow R_1 + \frac{1}{2}R_2$. The second operation is a column operation in which $\frac{1}{2}C_2$ is added to C_1 written as $C_1 \leftarrow C_1 + \frac{1}{2}C_2$.

$$\begin{pmatrix} 1 & 2 & 3 \\ 4 & 5 & 6 \\ 7 & 8 & 9 \end{pmatrix} \rightarrow \begin{pmatrix} 3 & \frac{9}{2} & 6 \\ 4 & 5 & 6 \\ 7 & 8 & 9 \end{pmatrix} \qquad \begin{pmatrix} 1 & 2 & 3 \\ 4 & 5 & 6 \\ 7 & 8 & 9 \end{pmatrix} \rightarrow \begin{pmatrix} 2 & 2 & 3 \\ \frac{13}{2} & 5 & 6 \\ 11 & 8 & 9 \end{pmatrix}. \qquad (2.9)$$

Below are the elementary type III matrices that correspond to these operations.

$$E_{R_1 \leftarrow R_1 + \frac{1}{2} R_2} = \begin{pmatrix} 1 & \frac{1}{2} & 0 \\ 0 & 1 & 0 \\ 0 & 0 & 1 \end{pmatrix} \qquad E_{C_1 \leftarrow C_1 + \frac{1}{2} C_2} = \begin{pmatrix} 1 & 0 & 0 \\ \frac{1}{2} & 1 & 0 \\ 0 & 0 & 1 \end{pmatrix}.$$

There is a very nice relationship between the elementary *operations* and elementary *matrices* that we state below without proof.

We can perform an elementary row operation r on M by forming the product $E_r M$, where E_r is the elementary matrix associated with r. The product $M E_c$ gives the corresponding elementary column operations c on M.

To illustrate, the operation $R_1 \leftarrow R_1 + \frac{1}{2} R_2$ in (2.9) can be computed as

$$\begin{pmatrix} 1 & \frac{1}{2} & 0 \\ 0 & 1 & 0 \\ 0 & 0 & 1 \end{pmatrix} \begin{pmatrix} 1 & 2 & 3 \\ 4 & 5 & 6 \\ 7 & 8 & 9 \end{pmatrix} = \begin{pmatrix} 3 & \frac{9}{2} & 6 \\ 4 & 5 & 6 \\ 7 & 8 & 9 \end{pmatrix}.$$

To obtain $C_1 \leftarrow C_1 + \frac{1}{2} C_2$, we have

$$\begin{pmatrix} 1 & 2 & 3 \\ 4 & 5 & 6 \\ 7 & 8 & 9 \end{pmatrix} \begin{pmatrix} 1 & 0 & 0 \\ \frac{1}{2} & 1 & 0 \\ 0 & 0 & 1 \end{pmatrix} = \begin{pmatrix} 2 & 2 & 3 \\ \frac{13}{2} & 5 & 6 \\ 11 & 8 & 9 \end{pmatrix}.$$

A common idea among our diagonalization algorithms is to perform the same row and column operations on a symmetric matrix M. This amounts to

$$E_{r_k} \dots E_{r_2} \ E_{r_1} \ M \ E_{c_1} \ E_{c_2} \dots E_{c_k}. \qquad (2.10)$$

We call such pairs of operations *congruence operations*. It is interesting to note that elementary type I and type II matrices are symmetric and $E_r = E_c$, and therefore, $E_r^T = E_c$. A type III elementary matrix is not symmetric yet $E_r^T = E_c$. The elementary matrices are nonsingular. Using Eq. (2.5), we see that (2.10) is of the form $P^T M P$ for a nonsingular P and thus congruent to M.

2.5   Spectral Graph Theory

Spectral graph theory seeks to describe graph properties with the eigenvalues and eigenvectors of the matrices associated with them. In this section, we define the adjacency and Laplacian matrices, historically the first to be used in spectral graph theory. Other matrices have been studied such as the normalized Laplacian matrix [9], signless Laplacian matrix [10], and distance matrix. According to [1], the distance matrix was already known to Cayley in the 1800s and was first formally studied in the 1930s, but its spectral study started in the 1970s.

Adjacency Matrix Let $G = (V, E)$ be a graph of order n, where $V = \{v_1, v_2, \ldots, v_n\}$. The *adjacency matrix* $A = A(G)$ of G is the $n \times n$ matrix of zeros and ones in which $A_{ij} = 1$ if $\{v_i, v_j\} \in E$ and is 0 otherwise. The adjacency matrix of the graph in Fig. 2.1 is

$$\begin{pmatrix} 0 & 1 & 1 & 0 & 0 \\ 1 & 0 & 0 & 1 & 0 \\ 1 & 0 & 0 & 1 & 1 \\ 0 & 1 & 1 & 0 & 1 \\ 0 & 0 & 1 & 1 & 0 \end{pmatrix}.$$

This matrix was first used for graph representation. It clearly contains all the information of the graph's structure. Historically, $A(G)$ was the first matrix to be studied in spectral graph theory.

One might wonder if the ordering of the vertices in a graph can affect the eigenvalues of the matrix. Fortunately, the answer is no. Consider two orderings of the vertices producing adjacency matrices $A_1 = A_1(G)$ and $A_2 = A_2(G)$. Since any permutation is a product of transpositions, the columns and rows of A_1 can be permuted by a sequence of row and column interchanges to obtain A_2. Each interchange may be obtained by multiplying an elementary type I matrix in the following way:

$$A_2 = E_k \ldots E_1 A_1 E_1 \ldots E_k.$$

Moreover, we have $E_i = E_i^{-1}$ for each i. So by (2.3), we have $A_2 = P^{-1} A_1 P$. Therefore, A_2 and A_1 are similar. By Theorem 2.1, they have the same eigenvalues.

This argument shows that isomorphic graphs have the same spectrum, as an isomorphism is simply a relabeling of the vertex set. Since there is no known efficient method for deciding whether two graphs are isomorphic, and given the nice fact that eigenvalues may be approximated efficiently, one of the early applications of spectral graph theory was in checking whether graphs are isomorphic or not. In an ideal world, we would be able to test graph isomorphism by simply computing the eigenvalues of both graphs. However, it is not true that two non-isomorphic graphs must have distinct spectra, and a graph G may have a *cospectral mate* H, namely a graph that is not isomorphic to G, but has the same spectrum as G.

Since $A(G)$ is a symmetric real matrix, its eigenvalues are real numbers. We number the eigenvalues of an adjacency matrix as follows:

$$\lambda_1 \geq \lambda_2 \geq \ldots \geq \lambda_n.$$

If λ is an eigenvalue of $A(G)$ with eigenvector $(x_1, \ldots, x_n)^T$, one sees that for each vertex v_i,

$$\sum_{v_j \in N(v_i)} x_j = \lambda x_i.$$

Here are some facts relating the structure of G to the spectrum of $A(G)$. The proofs of (a) and (c) can be found in [7], and (b) appears in [22].

(a) A graph is bipartite if and only if the spectrum of $A(G)$ is symmetric about the origin.
(b) The eigenvalue λ_1 satisfies $\hat{d} \leq \lambda_1 \leq \Delta$, where $\hat{d}$ is the average degree and Δ is the maximum degree of the graph.
(c) In a connected graph, the number of distinct eigenvalues of $A(G)$ is at least $diam(G) + 1$.

In 1967, Wilf [25] discovered a beautiful upper bound for the chromatic number of a graph, which states that $\chi(G) \leq 1 + \lambda_1$ for connected graphs. Equality happens if and only if $G = K_n$ or $G = C_k$ for k odd. It is easy to show that $\chi(G) \leq 1 + \Delta(G)$ using a greedy coloring algorithm. Since $\lambda_1 \leq \Delta$, Wilf's upper bound is better. His theorem generalizes the 1941 result known as Brooks' Theorem [6], a classical theorem in graph theory that says that $\chi(G) \leq \Delta(G) + 1$ with equality holding if and only if G is a complete graph or odd cycle.

Interlacing theorems relate the eigenvalues of a graph G and a subgraph H, where H is obtained by a certain operation such as deleting a vertex or an edge from G. Here is one example. Other interlacing results can be found in [16].

Theorem 2.2 *Consider a graph* $G = (V, E)$, *and let* $H = G - v$, *where* $v \in V$. *Let the eigenvalues of* $A(G)$ *and* $A(H)$ *be, respectively,*

$$\lambda_1 \geq \lambda_2 \geq \ldots \geq \lambda_{n-1} \geq \lambda_n$$

$$\theta_1 \geq \theta_2 \geq \ldots \geq \theta_{n-1}.$$

Then $\lambda_i \geq \theta_i \geq \lambda_{i+1}$, *for all* $1 \leq i \leq n - 1$.

Laplacian Matrix The *Laplacian matrix* of G is defined as follows. Let D be the diagonal matrix in which $D_{ii} = \deg(v_i)$. Then $L(G) = D - A(G)$. The Laplacian matrix of the graph in Fig. 2.1 is

$$
\begin{pmatrix}
2 & -1 & -1 & 0 & 0 \\
-1 & 2 & 0 & -1 & 0 \\
-1 & 0 & 3 & -1 & -1 \\
0 & -1 & -1 & 3 & -1 \\
0 & 0 & -1 & -1 & 2
\end{pmatrix}.
$$

The matrix $L(G)$ is also symmetric and has real eigenvalues. It is known that these eigenvalues always lie in $[0, n]$ and 0 is an eigenvalue. We number them as follows:

$$
\mu_1 \geq \mu_2 \geq \ldots \geq \mu_n = 0.
$$

If μ is an eigenvalue of $L(G)$ with eigenvector $(x_1, \ldots, x_n)^T$, one sees that for each vertex v_i, we have

$$
\deg(v_i)x_i \;-\; \sum_{v_j \in N(v_i)} x_j = \mu x_i.
$$

As early as 1973, Fiedler [12] discovered some striking properties of this matrix and showed the graph is connected if and only if $\mu_{n-1} > 0$. The eigenvalue μ_{n-1} is called the *algebraic connectivity* of G. The surveys [14, 15, 21] contain more results on the Laplacian spectrum.

The eigenvalues of $A(G)$ or $L(G)$ are roots of characteristic polynomials of the form $x^n + a_1 x^{n-1} + \ldots + a_{n-1}x + a_n$, where the a_is are integers. The roots of such polynomials are algebraic integers, and since they are real, we claim they must be irrational or integer. Suppose $\frac{p}{q}$ is a rational root, and assume p and q are relatively prime integers. This leads to

$$
p^n = -a_1 p^{n-1} q \ldots - a_{n-1} p q^{n-1} - a_n q^n. \tag{2.11}
$$

Since q divides the right side of (2.11), $q | p^n$ concluding that $q | p$ and $q = 1$, given that p and q are relatively prime.

If I is a real interval and M is a matrix, then $m_M I$ denotes the number of eigenvalues, multiplicities included, of M in I. If it is clear that we are talking about adjacency matrices, we can write, for example, $m_G[0, 1)$ instead of $m_{A(G)}[0, 1)$. Our intervals need not be finite; for example, we allow $m_M[1, \infty)$ and $m_M(-\infty, 3)$.

Fix a matrix $M(G)$ associated with graphs. An algorithm is said to be an *eigenvalue location algorithm* for a class of graphs $\mathcal{G}$, if for any $G \in \mathcal{G}$ and real interval I, it returns $m_{M(G)}I$.

Symmetric Matrix with Underlying Graph Given an $n \times n$ symmetric matrix $M = (M_{ij})$, we may associate it with a graph $G = (V, E)$, where $V = \{1, \ldots, n\}$

Fig. 2.4 A weighted graph of
order 5

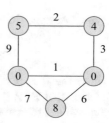

and $\{i, j\} \in E$ if and only if $M_{ij} \neq 0$ and $i \neq j$. We say that G is the *underlying graph* of M. It is easy to see that the matrix

$$M = \begin{pmatrix} 5 & 2 & 9 & 0 & 0 \\ 2 & 4 & 0 & 3 & 0 \\ 9 & 0 & 0 & 1 & 7 \\ 0 & 3 & 1 & 0 & 6 \\ 0 & 0 & 7 & 6 & 8 \end{pmatrix}$$

is symmetric matrix and its underlying graph appears in Fig. 2.1. In fact, this graph may be viewed as a *weighted graph* such that each edge $\{i, j\}$ has weight $M_{ij} \neq 0$ and each vertex i has weight M_{ii}, as shown in Fig. 2.4. The symbol $S(G)$ often denotes the set of symmetric matrices whose underlying graph is G.

2.6   Sylvester's Law of Inertia

If a real symmetric $n \times n$ matrix M has p positive eigenvalues, q negative eigenvalues, and r zero eigenvalues, then we will say M has *inertia* (p, q, r). Note $p, q, r \geq 0$ and $p + q + r = n$. The following result is in [4, Thr. 7.2].

Lemma 2.1 *If M is a real symmetric matrix, then there is an orthogonal matrix P such that $P^T M P$ is a diagonal matrix D.*

   Orthogonal means that $P^T = P^{-1}$, and so M is similar to a diagonal matrix. By Theorem 2.1, the diagonal entries are its eigenvalues. Recall that A and B are congruent if there exists a nonsingular P such that $A = P^T B P$. In the following, if $p = 0$, then I_p is empty. Similar remarks hold if $q = 0$ or $r = 0$.

Lemma 2.2 *If A is a real symmetric matrix with inertia (p, q, r), then*

$$A \cong \begin{pmatrix} I_p & & \\ & -I_q & \\ & & 0_r \end{pmatrix}. \qquad (2.12)$$

Proof Assume the eigenvalues of A are

$$\lambda_1, \lambda_2, \ldots, \lambda_p, -\lambda_{p+1}, \ldots, -\lambda_{p+q}, 0, \ldots, 0, \tag{2.13}$$

where the $\lambda_i > 0$, and zero has multiplicity r. By Lemma 2.1, there is a nonsingular P such that $P^T A P$ is a diagonal matrix D'. The diagonal matrix must contain the eigenvalues of A since $P^T = P^{-1}$. Without loss of generality, we may assume the diagonal elements are listed as in (2.13). Now let D be the diagonal matrix

$$\text{diag}\left(\frac{1}{\sqrt{\lambda_1}}, \ldots, \frac{1}{\sqrt{\lambda_p}}, \frac{1}{\sqrt{\lambda_{p+1}}}, \ldots, \frac{1}{\sqrt{\lambda_{p+q}}}, 1, \ldots, 1\right),$$

where r ones are listed. Let $C = PD$. Then C is nonsingular (as the product of two nonsingular matrices) and

$$C^T A C = D^T P^T A P D = D^T D' D = \begin{pmatrix} I_p & & \\ & -I_q & \\ & & 0_r \end{pmatrix},$$

showing A is congruent to the diagonal matrix in (2.12). $\square$

The following theorem is known as Sylvester's Law of Inertia, published in 1852 by J. J. Sylvester [23]. The proof below uses basic linear algebra properties. Another proof based on the continuity of eigenvalues can be found in [20].

Theorem 2.3 *Two $n \times n$ real symmetric matrices are congruent if and only if they have the same inertia.*

Proof First assume that A and B have the same inertia. By Lemma 2.2, both A and B are congruent to the matrix in (2.12). Since congruence is an equivalence relation, we must have $A \cong B$.

Now assume A and B are congruent, and A has inertia (p, q, r) and B has inertia (p', q', r'). Since A and B are congruent, they must be equivalent, and so their ranks are equal. Since the rank is the number of nonzero eigenvalues, this implies $r = r'$. By Lemma 2.2, we have

$$A \cong \begin{pmatrix} I_p & & \\ & -I_q & \\ & & 0_r \end{pmatrix} = E$$

$$B \cong \begin{pmatrix} I_{p'} & & \\ & -I_{q'} & \\ & & 0_r \end{pmatrix} = F.$$

It suffices to show $p = p'$. Suppose by contradiction that $p > p'$. Since $A \cong B$, we must have $E \cong F$. Therefore, there exists a nonsingular K such that $F =$

$K^T EK$. Next, let X be the $n \times p'$ submatrix of K consisting of the first p' columns. And let Y be the $n \times (n - p')$ submatrix consisting of the last $n - p'$ columns. We can write $K = (X|Y)$. Now define two subspaces U and W of $\mathbb{R}^{n \times 1}$ as follows:

$$U = \{\mathbf{x} \mid \mathbf{x} = Y\mathbf{y}, \ \mathbf{y} \in \mathbb{R}^{(n-p') \times 1}\}$$
$$W = \{\mathbf{x} \in \mathbb{R}^n \mid \mathbf{x}^T = (x_1, \ldots, x_p, 0, \ldots, 0)\}.$$

The subspace U is called the *range space* of Y. It is known [20, p. 199] that $\dim(U) = \operatorname{rank}(Y) = n - p'$. It is clear that $\dim(W) = p$. From (2.2), we have

$$\dim(U \cap W) = \dim(U) + \dim(W) - \dim(U + W). \tag{2.14}$$

By the assumption, the right side of (2.14) is $p - p' + n - \dim(U + W) > 0$, and so there must be a nonzero vector $\mathbf{x} \in U \cap W$. If $\mathbf{x} \in U$,

$$\mathbf{x} = Y\mathbf{y} = K \begin{pmatrix} \mathbf{0} \\ \mathbf{y} \end{pmatrix}.$$

Therefore,

$$\mathbf{x}^T E \mathbf{x} = \begin{pmatrix} \mathbf{0} \\ \mathbf{y} \end{pmatrix}^T K^T E K \begin{pmatrix} \mathbf{0} \\ \mathbf{y} \end{pmatrix} = (\mathbf{0}^T \mathbf{y}^T) F \begin{pmatrix} \mathbf{0} \\ \mathbf{y} \end{pmatrix} \leq 0$$

since the zero part of the vector has length p'. If $\mathbf{x} \in W$, then $\mathbf{x} = (x_1, \ldots, x_p, 0, \ldots, 0)^T$ and $\mathbf{x}^T E \mathbf{x} > 0$, since $p > 0$ and some $x_i \neq 0$. This is a contradiction. There is an extreme case when $p' = 0$, but the above argument still applies.

On the one hand, suppose $p < p'$. Interchanging p and p' in the definitions of U and W leads to a similar contradiction. $\qquad\square$

Corollary 2.1 *For fixed n, congruence has only finitely many equivalence classes.*

2.7 Analysis of Algorithms

To operate on mathematical objects such as graphs, polynomials, or nonassociative algebras, there must be an *encoding scheme* that maps these objects into *strings*. If G is a graph, we could encode it by listing the 0s and 1s in its adjacency matrix.

Algorithm analysis is concerned with the resources *time* and *space*. That is, the running time of an algorithm and the space needed to represent its *data structures*. As one would expect, these quantities depend on the size n of the input string. Typically, they increase as n increases. These ideas are explored further in [5].

We often focus on time rather than space for the following reason. If an algorithm has efficient running time, then it must also be efficient in terms of space. The reason is that on any standard computing model such as the deterministic Turing machine and its variants, in one unit of running time, a program can be written to only a constant number of memory locations. As is observed in [19], space can be reused but time cannot.

Running time is typically measured by the number of arithmetic operations with the understanding that running the same algorithm on different computers will only affect the running time by a constant. What is essential is the *growth rate* of the running time as a function of the input size n. We do not care about small n, but rather how the algorithm performs as n gets large. We approximate these using O notation, as defined below. Typically, algorithm times will vary over all inputs of the same size, so to be safe, we are usually interested in *worst case*. Let $t(n) \colon \mathbb{N} \to \mathbb{N}$ denote the worst case running time on an input of size n, and $f(n) \colon \mathbb{N} \to \mathbb{R}$. To say that $t(n)$ is in $O(f(n))$ means that there exists a constant c such that for sufficiently large n, one has $t(n) \le cf(n)$.

Here are two simple examples. Consider ordinary addition where the input is two n-digit integers. We will estimate time by the number of arithmetic operations. On some inputs, exactly n arithmetic operations are required. In the worst case, there are n adds and n carry operations. So the algorithm runs in $O(n)$, which is often called *linear time*. Now consider multiplication of two n-digit numbers using the traditional algorithm that we were taught in school (although there exist faster algorithms). This is an $O(n^2)$ algorithm.

There is a consensus among computer scientists that a good definition of an *efficient* running time is *polynomial time,* that is $O(n^k)$ for some k. However, there can be a huge difference between $O(n)$ and $O(n^3)$. Consider two algorithms A and B for the same problem where A is linear time and B is cubic. Disregarding the hidden constant, A can examine instances of size $n = 1000$ in the same time that B can examine sizes of $n = 10$. With a single processor, $O(n)$ is the best possible running time, since merely reading the input takes that amount of time.

Some problems seem to be inherently hard. There is no known polynomial-time algorithm for computing the chromatic number of a graph, or even deciding whether it is at most three. Computing a graph's domination number, independence number, and clique number also seem *intractable* and are known to be NP-hard [13].

2.8 Rooted Trees

A *rooted tree* is a modified tree that is often used for algorithms and can be constructed with the following process. For any tree $T = (V, E)$, choose an arbitrary node $r \in V$ as the tree's *root*. Each neighbor of r is regarded as a *child* of r, and r is called its *parent*. For each child c of r, all of its neighbors, except the parent, become its children. This process continues until all nodes except r have parents. A node that does not have children is called a *leaf* of the rooted tree. The

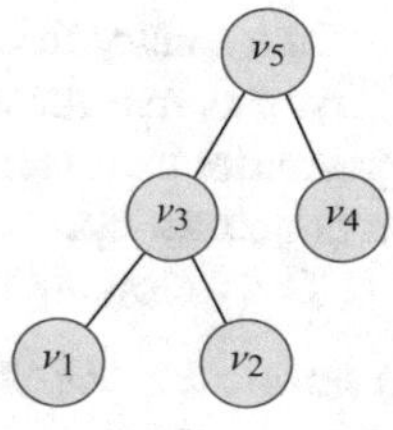

Fig. 2.5 A rooted tree with
root v_5

Table 2.1 A rooted tree
representation

Node	Children	
1		
2		
3	1	2
4		
5	3	4

remaining nodes are called *internal nodes* of the rooted tree. Figure 2.5 depicts a
rooted tree whose root is v_5 and has children v_3 and v_4. The children of v_3 are v_1
and v_2. The leaves are v_1, v_2, and v_4. In a *binary tree*, every internal node has two
children, known as the *left child* and the *right child*. The tree in Fig. 2.5 is a binary
tree. The *depth* of a node in a rooted tree is its distance to the root. The *depth of a
rooted tree* is the maximum depth of a node.

The *descendants* of a vertex in a rooted tree are defined as follows. If v is a leaf,
then it has no descendants. Otherwise, its descendants are its children, together with
their descendants. Given a rooted tree T and vertex v, the *subtree T_v* rooted at v is
the rooted tree consisting of v and all its descendants. If w is a descendant of v, we
say v is an *ancestor* of w.

If v is a leaf, there is a unique path from v to the root, which visits only the
ancestors of v. The *lowest common ancestor* of two leaves u and v is the node of
greatest depth that lies on both paths. This concept will be used throughout this
book. In Fig. 2.5, the lowest common ancestor of v_1 and v_2 is v_3, and the lowest
common ancestor of v_1 and v_4 is v_5.

In a *bottom-up* algorithm on a rooted tree, the children of a node must be
processed before the node can be processed. Therefore, we process the leaves first
and then the root last. Now consider a data structure that can represent a rooted tree
in linear space. Table 2.1 shows how the rooted tree in Fig. 2.5 could be represented.
Since trees of order n have exactly $n-1$ edges, we can represent the rooted tree with
an input of size $O(n)$.

A *traversal* is an algorithm that visits every vertex in a graph. A *depth-first
traversal* in a rooted tree T can be defined recursively as follows. First visit the
root r of the tree. For each child c of r, perform a depth-first traversal on the subtree
T_c rooted at c. A depth-first traversal on the tree in Fig. 2.5 would visit the vertices
as follows: v_5, v_3, v_1, v_2, and v_4. A depth-first traversal can also be done on graphs
if we keep track of the vertices visited.

In a binary tree T with root r, a *postorder traversal* can be defined as follows. First traverse the left subtree, then traverse the right subtree, and then visit r. A postorder traversal on the tree in Fig. 2.5 would visit v_1, v_2, v_3, v_4 and then v_5. Note it is bottom-up.

The following lemma will be used in Chap. 5.

Lemma 2.3 *Let T be any rooted tree. For each internal node v in T, let c_v denote its number of children. Then the number of leaves $\ell(T)$ in T is given by $\ell(T) = 1 + \sum_v (c_v - 1)$ where the sum is taken over all internal nodes v.*

Proof We prove this by induction on k, the number of internal nodes in T. When $k = 1$, the tree consists of a root r and its c_r children, which are all leaves. Then $\ell(T) = 1 + (c_r - 1)$. Now assume the formula holds for trees having $k > 1$ internal nodes. Let T be a tree with $k + 1$ internal nodes. Choose an internal node v in T of greatest depth. It has exactly c_v children, but they are all leaves. Remove the children of v forming a tree T' with k internal nodes. In the tree T', v is now a leaf. By the induction assumption,

$$\ell(T') = 1 + \sum_{v'} (c_{v'} - 1),$$

where v' denotes the internal nodes of T'. Since v is not a leaf in T, we have $\ell(T) = \ell(T') - 1 + c_v$. This gives

$$\ell(T) = \ell(T') - 1 + c_v = 1 + \sum_{v'} (c_{v'} - 1) + (c_v - 1).$$

The internal nodes in T are precisely the internal nodes in T' together with v. So we have $\ell(T) = 1 + \sum_v (c_v - 1)$, completing the induction.   □

References

1. Aouchiche, M., Hansen, P.: Distance spectra of graphs: A survey. Linear Algebra Appl. **458**, 301–386 (2014)
2. Bollobás, B.: Modern Graph Theory, 1 edn. Graduate Texts in Mathematics 184. Springer, New York (1998)
3. Bondy, J., Murty, U.: Graph Theory, 1st edn. Springer (2008)
4. Bradley, G.L.: A Primer of Linear Algebra. Prentice Hall, Englewood Cliffs, NJ (1973)
5. Brassard, G., Bratley, P.: Algorithmics. Prentice Hall, Englewood Cliffs, NJ (1988)
6. Brooks, R.L.: On coloring the nodes of a network. Proc. Cambridge Philos. Soc. **37**, 194–197 (1941)
7. Brouwer, A.E., Haemers, W.: Spectra of Graphs. Springer (2012)
8. Chartrand, G., Zhang, P.: A First Course in Graph Theory. Dover, Mineola, NY (2012)
9. Chung, F.R.K.: Spectral Graph Theory. American Mathematical Society (1997)
10. Cvetković, D., Doob, M., Sachs, H.: Spectra of Graphs, 3rd edn. Johann Ambrosius Barth Verlag, Heidelberg-Leipzig (1995)

11. Diestel, R.: Graph Theory, 5th edn. Springer (2017)
12. Fiedler, M.: Algebraic connectivity of graphs. Czechoslovak Math. J. **23(98)**, 298–305 (1973)
13. Garey, M.R., Johnson, D.S.: Computers and intractability. W. H. Freeman and Co., San Francisco, Calif. (1979). A guide to the theory of NP-completeness, A Series of Books in the Mathematical Sciences
14. Grone, R., Merris, R.: The Laplacian spectrum of a graph. II. SIAM J. Discrete Math. **7**(2), 221–229 (1994)
15. Grone, R., Merris, R., Sunder, V.S.: The Laplacian spectrum of a graph. SIAM J. Matrix Anal. Appl. **11**(2), 218–238 (1990)
16. Hall, F.J., Patel, K., Stewart, M.: Interlacing results on matrices associated with graphs. J. Combin. Math. Combin. Comput. **68**, 113–127 (2009)
17. Harary, F.: Graph Theory. Addison Wesley, Reading, MA (1969)
18. Horn, R.A., Johnson, C.R.: Matrix Analysis, 2nd edn. Cambridge University Press, New York, NY, USA (2012)
19. Kleinberg, J., Tardos, E.: Algorithm Design. Addison Wesley (2005)
20. Meyer, C.: Matrix analysis and applied linear algebra. Society for Industrial and Applied Mathematics (SIAM), Philadelphia, PA (2000)
21. Mohar, B.: Laplace eigenvalues of graphs—a survey. Discrete Math. **109**(1–3), 171–183 (1992). Algebraic graph theory (Leibnitz, 1989)
22. Stevanović, D.: Spectral radius of graphs. Elsevier Science, Saint Louis (2015)
23. Sylvester, J.J.: A demonstration of the theorem that every homogeneous quadratic polynomial is reducible by real orthogonal substitutions to the form of a sum of positive and negative squares. Philosophical Magazine. 4th Series **4**(23), 138–142 (1852)
24. West, D.B.: Introduction to Graph Theory, 2nd edn. Prentice Hall, New Delhi (2006)
25. Wilf, H.S.: The eigenvalues of a graph and its chromatic number. J. London Math. Soc. **42**, 330–332 (1967)

Chapter 3
Locating Eigenvalues in Trees

3.1 Adjacency Matrix

In this section, we describe the eigenvalue location algorithm for the adjacency matrix of trees that first appeared in [11]. The algorithm depends on the following corollary to Sylvester's Law of Inertia.

Corollary 3.1 *Let D be a diagonal matrix having p positive, q negative, and r zero entries. If M is real symmetric such that $M - \alpha I \cong D$, then M has:*

(a) p eigenvalues greater than α.
(b) q eigenvalues less than α.
(c) r is the multiplicity of the eigenvalue α.

Proof Since $M - \alpha I \cong D$, by Theorem 2.3, they have the same inertia. Since D has inertia (p, q, r), so does $M - \alpha I$. However, λ is an eigenvalue of M if and only if $\lambda - \alpha$ is an eigenvalue of $M - \alpha I$. If p of the values $\lambda - \alpha$ are positive, then there p eigenvalues of M greater than α, establishing (a). The other cases are similar. □

For an adjacency matrix $A(T)$ of a tree T and $x \in \mathbb{R}$, we will find a diagonal matrix D congruent to $A(T) + xI$ for any x. Recall from Eq. (2.10) in Chap. 2 that performing a sequence of identical elementary row and column operations preserves congruence. These are called congruence operations. The row and column operations will be type III, meaning that a multiple of a row (column) is added to another row (column).

Assume that the tree is rooted at any vertex and represented as shown in Table 2.1 of Sect. 2.7, so that each child of a parent can be determined in constant time. We assume the vertices $v_1, \ldots, v_n$ have been ordered so that if v_i is a child of v_k then $i < k$. If we manage to perform congruence operations in a way that annihilates the off-diagonal ones in the adjacency matrix without introducing new nonzero values off the diagonal, we find the desired diagonal matrix D. We show that this is possible by working bottom-up in the rooted tree and partially diagonalizing the matrix, while nodes are processed.

Below is a partially diagonalized matrix. While the first four rows and columns are diagonalized, the last three rows and columns are not.

$$\begin{pmatrix} 3 & 0 & 0 & 0 & 0 & 0 & 0 \\ 0 & 4 & 0 & 0 & 0 & 0 & 0 \\ 0 & 0 & 5 & 0 & 0 & 0 & 0 \\ 0 & 0 & 0 & 6 & 0 & 0 & 0 \\ 0 & 0 & 0 & 0 & 2 & 1 & 1 \\ 0 & 0 & 0 & 0 & 1 & 2 & 1 \\ 0 & 0 & 0 & 0 & 1 & 1 & 2 \end{pmatrix}.$$

We say a vertex v has been *processed* if the submatrix corresponding to v and its descendants is diagonal. Initially, all leaves are assumed to have been processed. In the above example, the vertex corresponding to the fifth row has been processed, but vertices corresponding to the sixth and seventh rows have not.

Consider an unprocessed parent v_k with diagonal value d_k all of whose children v_j have been processed, and we wish to process v_k. First assume all children have diagonal values $d_j \neq 0$. Consider a submatrix of the rows and columns for v_k and a child v_j. By performing the congruence operations

$$R_k \leftarrow R_k - \frac{1}{d_j} R_j \quad C_k \leftarrow C_k - \frac{1}{d_j} C_j,$$

the off-diagonal jk and kj entries will be annihilated as follows:

$$\begin{matrix} j \\ k \end{matrix} \begin{pmatrix} d_j & 1 \\ 1 & d_k \end{pmatrix} \longrightarrow \begin{matrix} j \\ k \end{matrix} \begin{pmatrix} d_j & 0 \\ 0 & d_k - 1/d_j \end{pmatrix}.$$

The *net effect* for all children is

$$d_k = d_k - \sum_c \frac{1}{d_c},$$

where c ranges over the children of v_k. Note that no new nonzero entries are created because the children have already been processed, and so there are only two nonzero values in the row and column of v_j.

On the other hand, suppose that $d_j = 0$ for some child v_j of v_k. Then we cannot divide by d_j. Select one such vertex v_j whose diagonal value is zero. We examine the submatrix corresponding to v_j and any sibling v_i, their parent v_k, and their grandparent v_ℓ, if present. As v_i and v_j are already processed, the operations

$$R_i \leftarrow R_i - R_j \quad C_i \leftarrow C_i - C_j$$

annihilate the two off-diagonal entries of any sibling v_i as follows:

$$
\begin{array}{c}
i \\ j \\ k \\ \ell
\end{array}
\left(
\begin{array}{cccc}
d_i & 1 & & \\
& 0 & 1 & \\
1 & 1 & d_k & 1 \\
& & 1 & d_\ell
\end{array}
\right)
\longrightarrow
\begin{array}{c}
i \\ j \\ k \\ \ell
\end{array}
\left(
\begin{array}{cccc}
d_i & 0 & & \\
& 0 & 1 & \\
0 & 1 & d_k & 1 \\
& & 1 & d_\ell
\end{array}
\right).
$$

If v_k is the root, we may ignore the next step. Otherwise, we can remove the two entries representing the edge between v_k and its parent v_ℓ with these operations:

$$
R_\ell \leftarrow R_\ell - R_j \quad C_\ell \leftarrow C_\ell - C_j.
$$

This transforms the submatrix as follows:

$$
\begin{array}{c}
i \\ j \\ k \\ \ell
\end{array}
\left(
\begin{array}{cccc}
d_i & 0 & & \\
& 0 & 1 & \\
0 & 1 & d_k & 1 \\
& & 1 & d_\ell
\end{array}
\right)
\longrightarrow
\begin{array}{c}
i \\ j \\ k \\ \ell
\end{array}
\left(
\begin{array}{cccc}
d_i & 0 & & \\
& 0 & 1 & \\
0 & 1 & d_k & 0 \\
& & 0 & d_\ell
\end{array}
\right).
$$

The following congruence operations

$$
R_k \leftarrow R_k - \frac{d_k}{2} R_j \quad C_k \leftarrow C_k - \frac{d_k}{2} C_j
$$

yield the transformation

$$
\begin{array}{c}
i \\ j \\ k \\ \ell
\end{array}
\left(
\begin{array}{cccc}
d_i & 0 & & \\
& 0 & 1 & \\
0 & 1 & d_k & 0 \\
& & 0 & d_\ell
\end{array}
\right)
\longrightarrow
\begin{array}{c}
i \\ j \\ k \\ \ell
\end{array}
\left(
\begin{array}{cccc}
d_i & 0 & & \\
& 0 & 1 & \\
0 & 1 & 0 & 0 \\
& & 0 & d_\ell
\end{array}
\right).
$$

Then the operations

$$
R_j \leftarrow R_j + R_k \quad C_j \leftarrow C_j + C_k
$$

transform as follows:

$$
\begin{array}{c}
i \\ j \\ k \\ \ell
\end{array}
\left(
\begin{array}{cccc}
d_i & 0 & & \\
& 0 & 1 & \\
0 & 1 & 0 & 0 \\
& & 0 & d_\ell
\end{array}
\right)
\longrightarrow
\begin{array}{c}
i \\ j \\ k \\ \ell
\end{array}
\left(
\begin{array}{cccc}
d_i & 0 & & \\
& 2 & 1 & \\
0 & 1 & 0 & 0 \\
& & 0 & d_\ell
\end{array}
\right).
$$

Finally,

$$R_k \leftarrow R_k - \frac{1}{2}R_j \quad C_k \leftarrow C_k - \frac{1}{2}C_j$$

yield the diagonalized form

$$
\begin{array}{c}
i \\ j \\ k \\ \ell
\end{array}
\left(
\begin{array}{cccc}
d_i & & 0 & \\
 & 2 & 1 & \\
0 & 1 & 0 & 0 \\
 & & 0 & d_\ell
\end{array}
\right)
\longrightarrow
\begin{array}{c}
i \\ j \\ k \\ \ell
\end{array}
\left(
\begin{array}{cccc}
d_i & & 0 & \\
 & 2 & 0 & \\
0 & 0 & -1/2 & 0 \\
 & & 0 & d_\ell
\end{array}
\right).
$$

The net effect is that

$$d_j = 2 \quad d_k = -\frac{1}{2}.$$

Furthermore, if v_k is not the root, the edge to its parent is removed.

This algorithm, `Diagonalize Tree`, is summarized in Fig. 3.1. Since rooted trees can be represented in $O(n)$ space, we see that the algorithm can be implemented in linear time. The algorithm needs only the rooted tree representation and an $O(n)$ array to store the diagonal values. It is *iterative* but can also be written *recursively*.

Recall that we use $m_M I$ to mean the number of eigenvalues of M in an interval I. If I is an infinite interval with endpoint a, then any eigenvalue location algorithm

```
Input: rooted tree T ordered from the bottom-up
Input: real number x
Output: diagonal matrix D = diag(d₁, ..., dₙ) congruent to A(T) + xI

Algorithm Diagonalize Tree(T, x)
   initialize dᵥ := x, for all vertices v
   for k = 1 to n
      if vₖ is a leaf then continue
      else if d_c ≠ 0 for all children c of vₖ then
          dₖ := dₖ - ∑ 1/d_c , summing over all children of vₖ
      else
          select one child vⱼ of vₖ for which dⱼ = 0
          dₖ := -½
          dⱼ := 2
          if vₖ has a parent vₗ, remove the edge {vₖvₗ}.
   end loop
```

Fig. 3.1 Diagonalizing $A(T) + xI$

that is based on Corollary 3.1 can determine $m_M I$ in one diagonalization. For example, diagonalizing with $x = -a$, we find $m_M(a, \infty)$ by the positive diagonal values returned, and $m_M[a, \infty)$ is the number of non-negative diagonal values. It should be clear that we can also deduce $m_M(-\infty, a)$ and $m_M(-\infty, a]$ from these diagonal values. When I is a bounded interval such as (a, b), we require two diagonalizations. Note that $m_M(a, b) = m_M(a, \infty) - m_M[b, \infty)$ and $m_M(a, b] = m_M(a, \infty) - m_M(b, \infty)$. The values $m_M[a, b)$ and $m_M[a, b]$ can be found similarly.

Eigenvalue Location Principle

Computing the number of eigenvalues in an infinite interval, I requires one diagonalization with $x = -a$. Computing the number of eigenvalues in a bounded interval, I requires two diagonalizations with $x = -a$ and $x = -b$.

Once we know that an eigenvalue resides in an interval $I = (a, b)$, we can determine which half of the interval by setting $x = -\frac{a+b}{2}$. This divide-and-conquer process can be repeated as many times as needed to approximate the eigenvalue.

To illustrate, consider diagonalizations of the same tree. As is customary, rooted trees are depicted with the root at the top and children placed beneath the parent. Consider the tree shown in part (a) of Fig. 3.2. We will show it has two eigenvalues less than -1 by diagonalization with $x = 1$. We use white to indicate that a node has been processed and gray to indicate unprocessed. Part (a) of the figure shows the

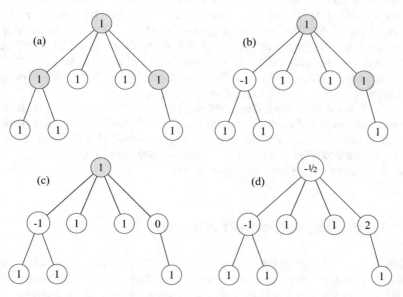

Fig. 3.2 Four steps in an application of the algorithm to show that $m_{A(T)}(-\infty, -1) = 2$; parts (a)–(d)

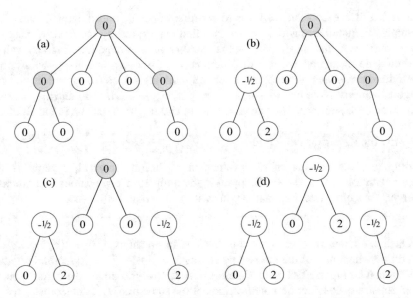

Fig. 3.3 Four steps in an application of the algorithm to compute the inertia of a tree; parts (a)–(d)

initialization and therefore only leaves have been processed. In part (b), a node with two children gets the value -1. In part (c), a node with a single child gets a zero. In the final step, part (d), the root has a child with zero. Consequently, the algorithm assigns 2 to the child and $-\frac{1}{2}$ to the root. No edge is broken because the root does not have a parent. We conclude that $m_{A(T)}(-\infty, -1) = 2$. By the symmetry of the spectrum of bipartite graphs, $m_{A(T)}(1, \infty) = 2$. If the algorithm had been performed with $x = 2$, we would obtain exactly one negative diagonal value, namely $-\frac{2}{3}$. This implies that $m_{A(T)}(-\infty, -2) = 1$. However from the previous computation, we must have $m_{A(T)}[-2, -1] = 1$ and also $m_{A(T)}(1, 2] = 1$.

In Fig. 3.3, we illustrate a diagonalization of this tree with $x = 0$. This really computes the inertia of the tree. We see that 0 has multiplicity two, and there are three positive eigenvalues and three negative eigenvalues. This would be expected since trees are bipartite, and so their spectra are symmetric about the origin. This example also illustrates edges getting removed.

3.2   Symmetric Matrices with Underlying Tree

A close look at the algorithm in Fig. 3.1 shows that the initial diagonal values can be arbitrary. In fact, the algorithm can be easily modified to locate eigenvalues in the Laplacian matrix of a tree by simply adding the vertex degrees to the diagonal during initialization. This algorithm first appeared in [5]. A second close look at Fig. 3.1

suggests that it is possible to allow arbitrary values in the off-diagonal positions if we multiply by the correct factors.

The adjacency matrix $A(T)$ and Laplacian matrix $L(T)$ are really special cases of the class of symmetric matrices whose underlying graph is a tree. A diagonalization algorithm by Braga and Rodrigues for this general matrix appears in [4] that we now describe.

Let $M = (m_{ij})$ be a symmetric $n \times n$ matrix whose underlying graph is a tree. This means the graph (V, E) defined by $V = \{1, \ldots, n\}$ where $\{i, j\} \in E$ if and only if $i \neq j$ and $m_{i,j} \neq 0$ is a tree T. Representing the matrix requires the structure of the tree T and the values in M. There are n diagonal values in M and, by symmetry, $n - 1$ off-diagonal values that must be considered. The diagonal values $m_{11}, \ldots, m_{nn}$ can be associated with the vertices, and the off-diagonal values can be associated with edges of T. We call this a *weighted* tree. The weighted tree can be represented in $O(n)$ space by augmenting the information in Table 2.1 with the weights.

Figure 3.4 shows `Diagonalize Weighted Tree`, a diagonalization algorithm for weighted trees. It is very similar to the algorithm for the adjacency matrix. The only differences are at initialization and multiplication by $(m_{jk})^2$ and $(m_{ck})^2$.

```
Input: weighted tree T ordered bottom-up, underlying M = (m_ij)
Input: real number x
Output: diagonal matrix D = diag(d_1, ..., d_n) congruent to M + xI

Algorithm Diagonalize Weighted Tree(T, x)
    initialize d_i := m_ii + x, for all i
    for  k = 1 to n
        if v_k is a leaf then continue
        else if d_c ≠ 0 for all children c of v_k then
            d_k := d_k - Σ (m_ck)²/d_c , summing over all children of v_k
        else
            select one child v_j of v_k for which d_j = 0
            d_k := - (m_jk)²/2
            d_j := 2
            if v_k has a parent v_l, remove the edge {v_k v_l}.
    end loop
```

Fig. 3.4 Diagonalizing $M + xI$ for a symmetric matrix M with an underlying tree

3.3   Laplacian Matrix and Applications

Assume that M is a Laplacian matrix $L(T) = D - A(T)$ in Fig. 3.4. Then we must have $m_{jk} = m_{ck} = -1$, which square to 1. Thus to diagonalize $L(T)$, we need to add only the vertex degrees to the diagonal at initialization, and then the process behaves *exactly* like the adjacency matrix diagonalization.

Figure 3.5 shows a Laplacian diagonalization on the path P_5 where $x = -2$. Note that in part (a) each vertex v initially has value $\deg(v) - 2$. After diagonalization in part (e), there are three negative values And two positive values. Hence, this graph has three Laplacian eigenvalues less than 2 and two Laplacian eigenvalues greater than 2. In fact, the Laplacian eigenvalues of P_n are known to be [7] in increasing order, $2 - 2\cos(i\pi/n)$, $i = 0, \ldots, n - 1$. Approximating to two decimal places, the Laplacian spectrum of P_5 is: 0, .38, 1.38, 2.61, 3.61.

In any graph with m edges, it is easy to see that the sum of the vertex degrees is $2m$, and therefore, the average degree is $\frac{2m}{n}$. Since trees have $m = n - 1$ edges, their average degree is $2m/n = 2(n-1)/n = 2 - 2/n$. It was conjectured in [18] that for any tree, at least half of the Laplacian eigenvalues are less than the average degree. Recently, this conjecture was settled independently by Sin [17] and the authors in [12].

Theorem 3.1 *For any tree T, $m_{L(T)}[0, 2 - \frac{2}{n}) \geq \lceil \frac{n}{2} \rceil$.*

The Laplacian diagonalization algorithm described above was crucial in both proofs. The theorem is equivalent to showing that for any tree T of order n, diagonalizing with $x = -2 + \frac{2}{n}$ must produce at least $\lceil \frac{n}{2} \rceil$ negative diagonal values.

The proofs of Theorem 3.1 are somewhat long and deep. There is a much simpler argument that $m_{L(T)}[0, 2) \geq \lceil \frac{n}{2} \rceil$ for all trees T. Unfortunately, this was not enough to establish the conjecture as Laplacian eigenvalues can occur in $(2 - \frac{2}{n}, 2)$. It is interesting that $\lim_{n \to \infty} 2 - \frac{2}{n} = 2$. In a tree, the path $P = u_0 u_1 \ldots u_p$, $p \geq 1$, is said to be a *pendant path attached at u_0* if the vertices $u_1, \ldots, u_{p-1}$ all have degree two and u_p is a leaf. A π-*transformation* combines two distinct pendant paths attached at the same vertex into a longer pendant path, as shown in Fig. 3.6.

Fig. 3.5 Five steps in the Laplacian diagonalization of P_5 with $x = -2$; parts (a)–(e)

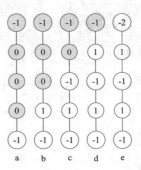

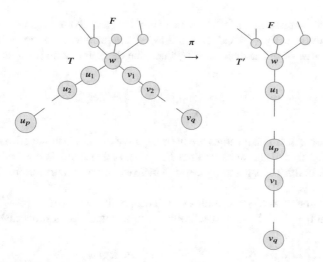

Fig. 3.6 π-transformation of T into T'

If we apply a π-transformation to T and obtain T', we can write $T \leq_\pi T'$. The following result by Mohar can be found in [13, Prop.2.1].

Lemma 3.1 *Every tree that is not a path has a vertex of degree at least three where at least two pendant paths are attached. Every tree can be transformed into a path by a sequence of π-transformations.*

In [5], Braga, Rodrigues and Trevisan show the interesting effect of π-transformations on the distribution of Laplacian eigenvalues. Note $m_{L(T)}[0, 2) \geq m_{L(T')}[0, 2)$ is equivalent to $m_{L(T)}[2, n] \leq m_{L(T')}[2, n]$ as there are n Laplacian eigenvalues in $[0, n]$.

Theorem 3.2 *If T is not a path and $T \leq_\pi T'$, then $m_{L(T)}[0, 2) \geq m_{L(T')}[0, 2)$.*

Proof We will show that after diagonalization of the two trees in Fig. 3.6 with $x = -2$, the tree T produces at least as many negative diagonal values as does T'. In both diagonalizations, we root the trees at w. The forest F obtained by removing w and the pendant paths will have the same diagonal values in both trees. Therefore, it is sufficient to consider only the diagonalization involving w and the pendant paths. We consider three cases depending on the parity of p and q.

First, assume p and q are both odd. When a pendant path is diagonalized with $x = -2$, as in Fig. 3.5, the diagonal values alternate $-1, 1, -1, 1, \ldots$ up to, but not including, the root. Therefore, the pendant paths in T will produce $\frac{p+1}{2} + \frac{q+1}{2} = \frac{p+q}{2} + 1$ negative values. The diagonal value for w is

$$d_w = \deg(w) - 2 - \left(\frac{1}{-1} + \frac{1}{-1} + \sum_{c \in C} \frac{1}{d_c} \right) = \deg(w) - \sum_{c \in C} \frac{1}{d_c},$$

where C is the set of children in the forest above the root. Now diagonalize T'. Since $p + q$ is even, exactly $\frac{p+q}{2}$ values equal to -1 are produced on the path. Note that in T' the degree of w has decreased by one. The diagonal value for w is

$$d'_w = \deg(w) - 1 - 2 - \left(\frac{1}{1} - \sum_{c \in C} \frac{1}{d_c}\right) = \deg(w) - 4 - \sum_{c \in C} \frac{1}{d_c}.$$

Clearly, $d'_w < d_w$. If these values are both negative or both non-negative, then T has one more negative diagonal value than T'. On the other hand, suppose that $d'_w < 0$ and $d_w \geq 0$. In this case, the two trees will have exactly the same number of negative diagonal values.

Now assume p and q are both even. Then both trees will have $\frac{p+q}{2}$ diagonal values equal to -1 in their pendant paths. Simple calculations show that

$$d_w = \deg(w) - 4 - \sum_{c \in C} \frac{1}{d_c} \qquad d'_w = \deg(w) - 4 - \sum_{c \in C} \frac{1}{d_c}.$$

Next assume that p is even and q is odd. Both trees have $\frac{p}{2} + \frac{q+1}{2}$ negative values in their pendant paths. Also,

$$d_w = d'_w = \deg(w) - 2 - \sum_{c \in C} \frac{1}{d_c}.$$

The case of p odd and q even is similar. This completes the proof.   □

Corollary 3.2 *For all trees T of order n, $m_{L(T)}[0, 2) \geq \lceil \frac{n}{2} \rceil$.*

Proof Let T be a tree of order n. By Lemma 3.1, T can be transformed into P_n by a sequence of π-transformations. Using Theorem 3.2 and induction, we must have $m_{L(T)}[0, 2) \geq m_{L(P_n)}[0, 2)$. It is known [5, Prop. 4.2] that $m_{L(P_n)}[0, 2 - \frac{2}{n}) = \lceil \frac{n}{2} \rceil$, and therefore, $m_{L(P_n)}[0, 2) \geq \lceil \frac{n}{2} \rceil$, implying $m_{L(T)}[0, 2) \geq \lceil \frac{n}{2} \rceil$.   □

In each of the diagonalization algorithms for trees, the value x is an eigenvalue when it produces a *permanent* zero diagonal value. In the case of Laplacian eigenvalues, suppose all children of v_k are nonzero, and suppose

$$\sum_c \frac{1}{d_c} = \deg(v_k) + x. \tag{3.1}$$

When v_k is processed, this produces a zero on the diagonal. If v_k is the root, then x is an eigenvalue. However, if v_k is not the root, the value could disappear and become 2 unless a sibling of v_k also has a zero.

Sin observes (see [17, Prop. 2.8]) that if (3.1) occurs, then there is *some* tree for which x is an eigenvalue. To prove this, he argues that if v_k is the root, then this is clear. However suppose that v_k is not the root. Then let T' be the rooted subtree consisting of v_k and all of its descendants. Let T_1 and T_2 be isomorphic copies of T' having roots r_1 and r_2, respectively. Now form a new rooted tree T'' by attaching a root r to r_1 and r_2. Applying the diagonalization algorithm to T'' with the same value x, the root r will have two children that become zero. One will stay permanently zero and the other will become 2. Therefore, x is an eigenvalue of T''.

In our experience, the benefit of eigenvalue location algorithms has usually not been in determining information about a particular graph. Rather the analysis of an algorithm can lead to interesting results about a class of graphs, as exemplified by Theorems 3.1 and 3.2.

We end the chapter by describing other applications of diagonalization and some open research problems that may be suitable for our diagonalization method. Still other applications of diagonalization in trees can be found in [6, 9, 15, 18].

One of the most celebrated problems in spectral graph theory, known as the *Brouwer Conjecture*, states that in any graph having m edges, the sum of the k largest Laplacian eigenvalues is at most $m + \frac{k(k-1)}{2}$. This had been verified for trees. In [8], the authors show, using diagonalization, that in any tree, the sum of the k largest Laplacian eigenvalues is at most $n - 1 + 2k - 1 - \frac{2k-2}{n}$, a better upper bound. If G is a graph with Laplacian eigenvalues $\mu_1, \ldots, \mu_{n-1}, \mu_n = 0$, its *Laplacian energy* is $\sum_1^n |\mu_i - \bar{d}|$, where $\bar{d}$ is the average degree. This bound enabled them to show that among all trees of order n, the star S_n has the largest Laplacian energy, settling an earlier conjecture by Radenković and Gutman [16]. It remains an open problem to characterize the n-vertex trees with the least Laplacian energy. It is conjectured that minimality is achieved by the path P_n.

The *index* of a graph, also called the *spectral radius*, is the largest eigenvalue of a matrix associated with it. If the matrix is omitted, index refers to the adjacency matrix. The method presented here has been successfully used to analyze the index of trees. This parameter is a very important graph invariant not only for pure mathematical reasons. Understanding how the spectral radius varies or finding the extremal values in a class of graphs has real-world application such as search engine problems [3] and the spread of diseases [19]. For more theoretical applications, we refer, for example, to the papers [1, 14] in which the authors were able to order certain classes of trees by the index of their adjacency matrices, enabling the solution of some hard problems in spectral graph theory. There still exist a great number of relevant classes of trees that would benefit from this kind of analysis.

Another line of investigation regarding the spectral radius for which eigenvalue location methods may be beneficial has to do with Hoffman–Smith (HS) limit points. For natural numbers $\Delta \geq 2$, the expression

$$\frac{\Delta}{\sqrt{\Delta - 1}},$$

is said to be a Hoffman–Smith limit point, inspired by the fact that Hoffman and Smith [10] considered the asymptotic behavior of the subdivision operator on the adjacency index ρ and the signless Laplacian index κ. Let G be a graph with maximum degree $\Delta(G) = \Delta$, and let $S^n(G)$ be the graph obtained from G by applying the subdivision operator n times.[1] Then

$$\lim_{n \to \infty} \rho(S^n(G)) = \frac{\Delta}{\sqrt{\Delta - 1}} \quad \text{and} \quad \lim_{n \to \infty} \kappa(S^n(G)) = \frac{\Delta^2}{\Delta - 1}.$$

For the adjacency matrix, the graphs whose index does not exceed the HS-limit 2 (HS-limit point for $\Delta = 2$) are known as *Smith graphs*. The HS-limit points for $\Delta = 3$ are $\frac{3}{2}\sqrt{2}$ for the adjacency matrix and $\frac{9}{2}$ for the signless Laplacian matrix. A complete characterization of the graphs whose ρ index is below $\frac{3}{2}\sqrt{2}$ or whose κ index is below $\frac{9}{2}$ is not known. This is in general a hard problem, but it has been possible to characterize classes of trees that satisfy this condition [2]. For other values of Δ, very little is known, and this is a wide open area of research.

References

1. Belardo, F., Oliveira, E.R., Trevisan, V.: Spectral ordering of trees with small index. Linear Algebra Appl. **575**, 250–272 (2019)
2. Belardo, F., Brunetti, M., Trevisan, V., Wang, J.: On quipus whose signless Laplacian index does not exceed 4.5. J. Algebraic Combin. (2022)
3. Berry, M.W. (ed.): Computational Information Retrieval. Society for Industrial and Applied Mathematics, USA (2001). https://doi.org/10.5555/762544
4. Braga, R., Rodrigues, V.: Locating eigenvalues of perturbed Laplacian matrices of trees. TEMA (São Carlos) Brazilian Soc. Appl. Math. Comp. **18**(3), 479–491 (2017)
5. Braga, R.O., Rodrigues, V.M., Trevisan, V.: On the distribution of Laplacian eigenvalues of trees. Discrete Math. **313**(21), 2382–2389 (2013)
6. Braga, R.O., Del-Vecchio, R.R., Rodrigues, V.M., Trevisan, V.: Trees with 4 or 5 distinct normalized Laplacian eigenvalues. Linear Algebra Appl. **471**, 615–635 (2015)
7. Brouwer, A.E., Haemers, W.: Spectra of Graphs. Springer (2012)
8. Fritscher, E., Hoppen, C., Rocha, I., Trevisan, V.: On the sum of the Laplacian eigenvalues of a tree. Linear Algebra Appl. **435**, 371–399 (2011)
9. Fritscher, E., Hoppen, C., Rocha, I., Trevisan, V.: Characterizing trees with large Laplacian energy. Linear Algebra Appl. **442**, 20–49 (2014). Special Issue on Spectral Graph Theory on the occasion of the Latin Ibero-American Spectral Graph Theory Workshop (Rio de Janeiro, 27–28 September 2012)
10. Hoffman, A.J., Smith, J.: On the spectral radii of topologically equivalent graphs. In: M. Fiedler (ed.) Recent Advances in Graph Theory, pp. 273–281. Academia, Prague (1975)
11. Jacobs, D.P., Trevisan, V.: Locating the eigenvalues of trees. Linear Algebra Appl. **434**(1), 81–88 (2011)
12. Jacobs, D.P., Oliveira, E., Trevisan, V.: Most Laplacian eigenvalues of a tree are small. J. Combin. Theory B **146**, 1–33 (2021)
13. Mohar, B.: On the Laplacian coefficients of acyclic graphs. Linear Algebra Appl. **722**, 736–741 (2007)

---

[1] This means all edges are subdivided, all edges of the resulting graph are subdivided, and this process is iterated n times.

14. Oliveira, E.R., Stevanović, D., Trevisan, V.: Spectral radius ordering of starlike trees. Linear Multilinear Algebra **68**(5), 991–1000 (2020)
15. Patuzzi, L., de Freitas, M.A.A., Del-Vecchio, R.R.: Indices for special classes of trees. Linear Algebra Appl. **442**, 106–114 (2014). Special Issue on Spectral Graph Theory on the occasion of the Latin Ibero-American Spectral Graph Theory Workshop (Rio de Janeiro, 27–28 September 2012)
16. Radenković, S., Gutman, I.: Total π-electron energy and Laplacian energy: how far the analogy goes? J. Serb. Chem. Soc. **72**(12), 1343–1350 (2007)
17. Sin, C.: On the number of Laplacian eigenvalues of trees less than the average degree. Discrete Math. **343**(10) (2020)
18. Trevisan, V., Carvalho, J.B., Del Vecchio, R.R., Vinagre, C.T.M.: Laplacian energy of diameter 3 trees. Appl. Math. Lett. **24**(6), 918–923 (2011)
19. Van Mieghem, P., Omic, J., Kooij, R.: Virus spread in networks. IEEE/ACM Trans. Networking **17**(1), 1–14 (2009). https://doi.org/10.1109/TNET.2008.925623

Chapter 4
Graph Classes and Graph Decompositions

4.1 Hereditary Graph Classes

Let $\mathcal{G}$ denote the set of all finite graphs. A *graph class C* is a subset of $\mathcal{G}$ that is closed under isomorphism. The set of connected graphs is one such class, as renaming vertices does not affect connectedness. Compare that with the set of graphs with a vertex labeled 1, which is not a graph class.

Given a fixed family $\mathcal{F}$ of graphs, we say that a graph G is *induced $\mathcal{F}$-free* (or simply *$\mathcal{F}$-free*) if it does not contain any element of $\mathcal{F}$ as an induced subgraph, that is, $F \not\vartriangleleft G$ for any $F \in \mathcal{F}$. Let **Forb**$(\mathcal{F})$ denote the class of all $\mathcal{F}$-free graphs, that is, the class of graphs for which induced copies of elements of $\mathcal{F}$ are *forbidden*. For instance, if $\mathcal{F}_1 = \{C_\ell : \ell \geq 3\}$ and $\mathcal{F}_2 = \{C_{2\ell+1} : \ell \geq 1\}$, then Forb$(\mathcal{F}_1)$ and Forb$(\mathcal{F}_2)$ are precisely the classes of forests and bipartite graphs, respectively.

A graph class C is said to be *hereditary* if it is closed under induced subgraphs. In other words, if $G \in C$ and H is an induced subgraph of G, then $H \in C$. Note that the class of connected graphs is not hereditary, while the classes of forests and bipartite graphs are both hereditary. The result below offers a characterization of hereditary graph classes.

Theorem 4.1 *A graph class C is hereditary if and only if $C = $ Forb$(\mathcal{F})$ for some graph family $\mathcal{F}$.*

Proof To show that Forb$(\mathcal{F})$ is hereditary, let $G \in$ Forb$(\mathcal{F})$, and fix an arbitrary induced subgraph $H \vartriangleleft G$. If $F \in \mathcal{F}$ and $F \vartriangleleft H$, then we get the contradiction $F \vartriangleleft G$ by transitivity of induced subgraphs. For the converse, let C be a hereditary graph class, and let $\mathcal{F} = \overline{C} = \mathcal{G} \setminus C$. By definition, if $G \notin C$, then $G \notin$ Forb$(\mathcal{F})$, so that Forb$(\mathcal{F}) \subseteq C$. Let $G \in C$. To show that $G \in$ Forb$(\mathcal{F})$, fix an arbitrary $H \vartriangleleft G$. Since C is hereditary, $H \in C = \overline{\mathcal{F}}$. Thus $C \subseteq$ Forb$(\mathcal{F})$, as required. $\square$

In the above proof, the family $\mathcal{F}$ may be chosen as the set of minimal elements of $\mathcal{G} \setminus C$, namely the graphs $F \in \mathcal{G} \setminus C$ such that $F \notin C$, but such that any proper

C. Hoppen et al., *Locating Eigenvalues in Graphs*, SpringerBriefs in Mathematics, https://doi.org/10.1007/978-3-031-11698-8_4

induced subgraph E of F satisfies $E \in C$. Finding such a minimal family explicitly is an important problem in the study of hereditary graph classes.

The hereditary class that has attracted the most attention is probably the class of *perfect graphs*, the class of graphs G with the property that, for any induced subgraph F of G, the chromatic number $\chi(F)$ and the clique number $\omega(F)$ coincide. Two major breakthroughs in the study of this class have been the proofs of the perfect graph conjecture by Lovász [28] and of the strong perfect graph conjecture by Chudnovsky et al. [10]. The latter establishes that a graph G is perfect if and only if it neither contains an odd cycle of length at least 5 (known as an *odd hole*) nor a complement of such an odd cycle (known as an *odd antihole*) as induced subgraphs.

Here, our focus will be on *cographs*, but we shall also mention other graph classes that are relevant in complexity theory, such as chordal graphs (see [25] for more examples). Moreover, the class of *distance-hereditary graphs* will be considered in depth in Chap. 8. There is a rich literature about hereditary graph classes that is well beyond the scope of this book, and we refer the interested reader to [9].

4.2   Cographs

In this section, we describe the class of complement reducible graphs, known as *cographs*, which is precisely the hereditary class of P_4-free graphs. This class appeared naturally in various contexts and has been rediscovered several times. The following is an excerpt from the seminal paper of Corneil, Lerchs, and Burlingham [12]:

> Cographs have arisen in many disparate areas of mathematics and have independently been rediscovered by various researchers. Names synonymous with cographs include D^*-graphs, P_4 restricted graphs, and HD or Hereditary Dacey graphs.

In fact, the versatility of this class is illustrated by the several equivalent ways in which its elements may be defined. One of the definitions uses *(disjoint) unions* and *complements*. Given graphs $G_1 = (V_1, E_1)$ and $G_2 = (V_2, E_2)$ whose vertex sets are disjoint, the union $G_1 \cup G_2$ is the graph with vertex set $V = V_1 \cup V_2$ and edge set $E = E_1 \cup E_2$. This can be naturally extended to the union $G_1 \cup \cdots \cup G_k$ of a finite family of mutually vertex-disjoint graphs $G_1 = (V_1, E_1), \ldots, G_k = (V_k, E_k)$. We omit parentheses in $G_1 \cup \cdots \cup G_k$ because the union is associative. It is easy to see that it is also commutative. The union of two graphs is depicted in Fig. 4.1. The authors of [12] recursively defined the class C of complement reducible graphs as follows:

(α) $K_1 \in C$.
(β) If G_1 and G_2 are vertex-disjoint graphs in C, then $G_1 \cup G_2 \in C$.
(γ) If $G \in C$, then $\overline{G} \in C$.

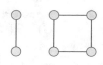
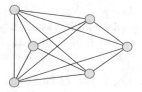

Fig. 4.1 The union $P_2 \cup C_4$ and the join $P_2 \oplus C_4$

A similar definition may be obtained through unions and *joins*, also known as *(disjoint) sums*. Given graphs $G_1 = (V_1, E_1)$ and $G_2 = (V_2, E_2)$ whose vertex sets are disjoint, the join $G_1 \oplus G_2$ is the graph with vertex set $V = V_1 \cup V_2$ and edge set $E = E_1 \cup E_2 \cup \{\{v_i, v_j\}: v_i \in V_1, v_j \in V_2\}$. This can be naturally extended to the join $G_1 \oplus \cdots \oplus G_k$ of a finite family of mutually vertex-disjoint graphs. We omit parentheses in $G_1 \oplus \cdots \oplus G_k$ because the join operation is associative (it is also commutative). Consider the class $\mathcal{D}$ defined recursively as follows:

(a) $K_1 \in \mathcal{D}$.
(b) If G_1 and G_2 lie in $\mathcal{D}$, then $G_1 \cup G_2 \in \mathcal{D}$.
(c) If G_1 and G_2 lie in $\mathcal{D}$, then $G_1 \oplus G_2 \in \mathcal{D}$.

Theorem 4.2 *If C and $\mathcal{D}$ are the above graph classes, then* $\mathrm{Forb}(P_4) = C = \mathcal{D}$.

Proof We first show that $\mathrm{Forb}(P_4) = C$. We start with $C \subseteq \mathrm{Forb}(P_4)$. Suppose that there is a minimum counterexample, a graph G that contains an induced copy of P_4 and is obtained from copies of K_1 through a shortest sequence of operations of type (β) and (γ). Let $n = |V(G)|$. Since $n \geq 4$, the last step in this shortest sequence is either $G = \overline{H}$ or $G = G_1 \cup G_2$, where H, G_1, and G_2 are P_4-free by the minimality of G. The first option cannot happen because $\overline{P_4} = P_4$, and therefore, H would contain an induced copy of P_4, contradicting the minimality of G. The second cannot happen because any induced copy of a connected graph would lie entirely within G_1 or within G_2, again contradicting the minimality of G.

To show that $\mathrm{Forb}(P_4) \subseteq C$, we use the following claim.

Claim If G is P_4-free and $|V(G)| \geq 2$, then G is disconnected or $\overline{G}$ is disconnected.

To see why this is true, assume for a contradiction that the statement fails for some P_4-free graph G, and fix one such counterexample with the minimum number of vertices n. This assumption implies that G and $\overline{G}$ are both connected. Because G is P_4-free if and only if $\overline{G}$ is P_4-free, $\overline{G}$ is also a minimal counterexample. We also have $n > 3$, as any connected graph with two or three vertices has a disconnected complement.

Let $x \in V(G)$ and consider the P_4-free graph $G - x$. Note that $\overline{G - x} = \overline{G} - x$, so that, by the minimality of G, we have $G - x$ or $\overline{G} - x$ disconnected. We may assume without loss of generality that $G - x$ is disconnected. Since $\overline{G}$ is connected, there is a vertex $y \in V(G)$ such that $\{x, y\} \notin E(G)$. Let C be the component of $G - x$ that contains y. Note that x has a neighbor z in C, as G is connected. In fact, by considering the vertices along the path P from z to y in C, we may fix such z and y with the additional property $\{z, y\} \in E(G)$. Since $G - x$ has more than one component and G is connected, x must have a neighbor $w \in V(G) \setminus V(C)$. This implies that $G[\{w, x, z, y\}]$ is isomorphic to P_4, a contradiction. □

With this, to show that $\mathrm{Forb}(P_4) \subseteq C$, we again assume that there is a P_4-free n-vertex graph G that is not in C with the property that any P_4-free graph with fewer than n vertices lies in C. We argue that G is connected. Indeed, if we suppose that G is disconnected, we have $G = G_1 \cup G_2$ for nonempty P_4-free graphs G_1, G_2. The minimality of G implies that $G_1, G_2 \in C$, leading to $G \in C$ by item (β) in the definition of C. This contradicts the choice of G, and therefore, G is connected. By the claim above, $\overline{G}$ is disconnected. Since $\overline{G}$ is P_4-free and its components are in C (again by the minimality of G), we conclude that $\overline{G} \in C$ by item (β) in the definition of C. This implies that $G \in C$ by item (γ), the desired contradiction.

The proof that $C = \mathcal{D}$ is easy. We prove that $C \subseteq \mathcal{D}$ by induction on the number n of vertices. Since $K_1 \in C \cap \mathcal{D}$, the result holds for graphs with $n = 1$. Assume that any graph in C with at most $n \geq 1$ vertices lies in $\mathcal{D}$, and fix $G \in C$ such that $|V(G)| = n + 1$. If $G = G_1 \cup G_2$ for some $G_1, G_2 \in C$, then $G \in \mathcal{D}$ by the induction hypothesis and part (b) of the definition of $\mathcal{D}$. Otherwise, $G = \overline{H}$, and we may assume that H is defined as a union, and not another complement. Note also that $H = \overline{G} \in C$. In particular, $\overline{G} = G_1 \cup G_2$, for some $G_1, G_2 \in C \cap \mathcal{D}$. This implies that $G = \overline{G_1} \oplus \overline{G_2}$, with $\overline{G_1}, \overline{G_2} \in C \cap \mathcal{D}$ by the induction hypothesis, thus $G \in \mathcal{D}$ as required. The proof that $\mathcal{D} \subseteq C$ follows from an easier induction argument, observing that $G_1 \oplus G_2 = \overline{\overline{G_1} \cup \overline{G_2}}$. □

Using the recursive construction corresponding to $\mathcal{D}$, we will be able to represent every cograph on vertex set $[n]$ as a rooted tree whose *nodes*[1] consist of n leaves labeled 1 through n and internal vertices that carry either the label "$\cup$" for union or "$\oplus$" for join. Given such a tree T, we construct the corresponding cograph G_T as follows: arbitrarily order the nodes of T as $1, \ldots, m$ in postorder (that is, they are ordered bottom-up). For i from 1 to m, process node i as follows. If it is a leaf labeled j, produce a singleton whose vertex is labeled j. If it is an internal node labeled $\cup$, take the union of the graphs produced by its children. If it is an internal node labeled $\oplus$, take the join of the graphs produced by its children. A tree T associated with a graph $G = G_T$ will be called a *cotree* of G.[2] A lot of information about a cograph may be easily obtained from a cotree that produces it. For instance, the next result tells us that it is easy to see whether two vertices of a graph G are adjacent based on a cotree of G. In particular, it implies that the cotree

---

[1] When referring to any auxiliary graph that records a structural decomposition of a graph, we use the word *node* instead of vertex to avoid ambiguity.

[2] The term cotree has been originally reserved to a similar construction where internal nodes are labeled by unions and complements, according to the recursive construction that defines C. However, it is common to also use this terminology for trees produced according to the recursive construction in $\mathcal{D}$, as we do here.

$\overline{T}$ obtained from T by switching labels $\oplus$ and $\cup$ produces the complement of G, that is, $G_{\overline{T}} = \overline{G}$.

Lemma 4.1 *Let $G = (V, E)$ be a graph, and let T be a cotree that produces it. Fix distinct vertices $u, v \in V$, and let x be the lowest common ancestor of the leaves corresponding to u and v in T. If x has label $\oplus$, then u and v are adjacent; if x has label $\cup$, then they are not.*

Proof We shall prove this result by induction on the *depth d* of the cotree. Since G has at least two vertices, T has at least two leaves and we have $d \geq 1$.

If $d = 1$, then the cotree T consists of a root r labeled $\cup$ or $\oplus$ connected to $|V(G)|$ leaves, each corresponding to a vertex of G. By definition of cograph, G is edgeless if r is labeled $\cup$, or is a complete graph if r is labeled $\oplus$. Also, node r is the lowest common ancestor of the leaves corresponding to any vertices u and v. Thus the lemma holds in this case.

Next suppose that the desired result holds for any cotree with depth at most $d \geq 1$, and let G, T, u, v, and x be as in the statement of the lemma, where T has depth $d + 1$. Let r be the root of T, and assume that it has $\ell \geq 1$ children $r_1, \ldots, r_\ell$. Suppose that r is labeled $\odot \in \{\cup, \oplus\}$. By definition of $G = G_T$, we have $G = G_1 \odot \cdots \odot G_\ell$, where each G_i is the cograph produced by the subtree T_i of T induced by r_i and its descendants (and rooted at r_i). Of course, each subtree has depth at most d. If $\ell = 1$, then $G = G_1$, and the result follows by induction using G and T_1, so suppose $\ell \geq 2$.

First assume that $u \in V(G_i)$ and $v \in V(G_j)$, where $i \neq j$. By the definition of $G = G_1 \odot \cdots \odot G_\ell$, this means that $x = r$ and that u and v are adjacent if and only if $\odot = \oplus$, as required. Next assume that $u, v \in V(G_i)$. This implies that u and v are adjacent in G if and only if they are adjacent in G_i. Moreover, the lowest common ancestor x of the leaves corresponding to u and v in T is a node of T_i. The required conclusion follows by induction. $\qquad\square$

Lemma 4.1 immediately leads to the following important property of cotrees.

Corollary 4.1 *Let $G = (V, E)$ be a graph, let T be a cotree that produces G, and let T' be a rooted subtree of T. The cotree T' produces the induced subgraph $G[S]$, where $S \subset V$ is the set of all vertices of G that are labels to leaves of T'.*

Assuming that we are given cotrees associated with cographs, it is possible to design linear time algorithms for problems that are generally assumed to be hard for general graphs. Such problems include computing graph parameters that have already been mentioned in this book, such as the chromatic number and the clique number, and applications to other classical problems, such as clustering, minimum weight domination, isomorphism, and Hamiltonicity, for instance. We illustrate this with the following example. Let $G = G_1 \oplus \cdots \oplus G_k$, where G_i is a cograph for each $i \in [k]$. So G is a cograph whose cotree is rooted at a node with label $\oplus$ and has branches given by the cotrees associated with $G_1, \ldots, G_k$. Then the clique number $\omega(G)$ satisfies $\omega(G) = \sum_{i \in [k]} \omega(G_i)$, while its independence number $\alpha(G)$ satisfies $\alpha(G) = \max_{i \in [k]} \alpha(G_i)$. Analogously, if $G = G_1 \cup \cdots \cup G_k$, we

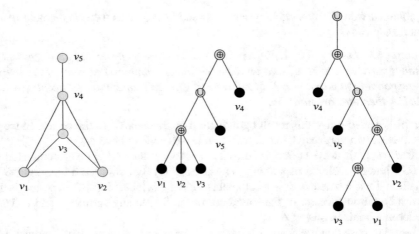

Fig. 4.2 A cograph and two cotrees associated with it

have $\omega(G) = \max_{i \in [k]} \omega(G_i)$ and $\alpha(G) = \sum_{i \in [k]} \alpha(G_i)$. This allows us to compute $\omega(G)$ and $\alpha(G)$ in linear time by recursively using the cotree of G. In fact, it gives us a strategy to compute a maximum clique or a maximum independent set in a cograph G. We explain this for independent sets. Consider a cotree associated with G and order its nodes bottom-up. Each leaf x is associated with a singleton I_x whose element is the vertex of G that labels it. At any other node x, define I_x as follows: if x is labeled $\cup$, it is the union of the sets I_y computed for its children; if x is labeled $\oplus$, $I_x = I_y$, where I_y has maximum size among all sets computed for the children of x.

As shown in Fig. 4.2, different cotrees may produce the same cograph. We say that a cotree is in *normalized form* if every internal node has at least two children and has a label that differs from the label of its parent. In other words, the children of nodes labeled $\cup$ are leaves or nodes labeled $\oplus$, while the children of nodes labeled $\oplus$ are leaves or nodes labeled $\cup$. One advantage of normalized cotrees is that they immediately give away whether the associated graph is connected. If the root of such a tree T is labeled $\cup$, then G_T is disconnected; otherwise, G_T is connected. Note that the left cotree in Fig. 4.2 is in normalized form. Figure 4.3 also depicts a cotree in normalized form.

Theorem 4.3 *Every cograph is associated with a cotree. The cotree is unique if it is in normalized form.*

Proof We prove the result by induction on the number of vertices n. We consider the definition of cograph given by the family $\mathcal{D}$.

For $n = 1$, the cograph $G = K_1$ with $V(G) = \{v\}$ is associated with a normalized cotree with a single node labeled v. Note that any other cotree T would contain at least two nodes and therefore at least two nodes with degree 1. Since G

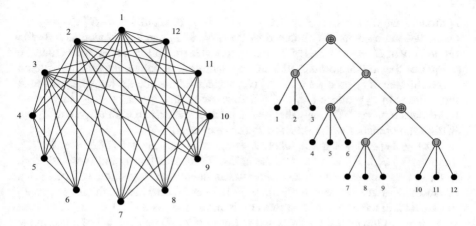

Fig. 4.3 A cograph and the corresponding cotree in normalized form

has a single vertex, one of the nodes with degree 1 must be labeled by $\cup$ or $\oplus$ and has a single child, and thus T is not in normalized form.

Next suppose that, for some $n \geq 1$, the statement holds for all cographs with at most n vertices, and let G be a cograph on $n+1 > 1$ vertices, so that $G = G_1 \cup G_2$ or $G = G_1 \oplus G_2$, where G_1 and G_2 are cographs with at most n vertices. By induction, there are normalized cotrees T_1 and T_2 associated with G_1 and G_2, respectively. Call their roots r_1 and r_2. Consider the cotree T obtained from T_1 and T_2 by introducing a new root node r and connecting it to r_1 and r_2 (r_1 and r_2 become the children of r in T). The root r is labeled $\cup$ if $G = G_1 \cup G_2$ and labeled $\oplus$ otherwise, and therefore, T is a cotree for G. We still need to get a cotree that is in normalized form.

For simplicity, assume that the label of r is $\cup$, and the argument is analogous in the other case. If neither r_1 nor r_2 is labeled $\cup$, the cotree T is in normalized form, and we are done. If r_1 is labeled $\cup$, we turn T into a new cotree T' by deleting r_1 (along with incident edges) and by connecting its children directly to r. We first claim that T' produces the same graph as T. Indeed, the leaves of T and T' are the same, and if we fix any two leaves, either their lowest common ancestor in both trees is the same or it is r_1 in T and r in T'. Since r_1 and r are labeled the same, our claim follows from Lemma 4.1. Moreover, because T_1 is normalized, no new children of r in T' are labeled $\cup$. If r_2 is labeled $\cup$, we do the same. The cotree redefined in this way produces G and is in normalized form.

It remains to show that the cotree is unique if it is in normalized form. We did this already for $n = 1$, and for a contradiction, assume that there is a minimal counterexample, namely a graph G with a minimum number of vertices $n \geq 2$ that is associated with two distinct normalized cotrees T_1 and T_2 rooted at r_1 and r_2, respectively. As we described above, there is a bijection between the (normalized) cotrees of G and of $\overline{G}$, so that G is a minimal counterexample if and only if $\overline{G}$ is a minimal counterexample. Thus we may assume that G is connected; hence, the roots

r_1 and r_2 have the same label $\oplus$. Let $G = G'_1 \oplus \cdots \oplus G'_k$ and $G = G''_1 \oplus \cdots \oplus G''_\ell$ be the decompositions of G induced by the roots of T_1 and T_2, respectively. By the minimality of G, the subtrees of T_1 and T_2 rooted at the children of r_1 and r_2 must be unique, as they are in normalized form. By definition of cotree, the branches rooted at the children of r_1 produce $G'_1, \ldots, G'_k$, respectively, while the branches rooted at the children of r_2 produce $G''_1, \ldots, G''_\ell$. Because of this, to get $T_1 \neq T_2$, there must be vertices $v, w \in V(G)$ that label leaves in the same branch with respect to T_1 but in different branches with respect to T_2 (or vice versa).

Fix $a \in [k]$ and $b, c \in [\ell]$, where $b \neq c$, such that $u, v \in V(G'_a)$, $u \in V(G''_b)$, and $v \in V(G''_c)$. By Lemma 4.1, the cotree T_2 tells us that u and v are adjacent in G, so that their lowest common ancestor in the normalized subtree T'_a of T_1 that produces G'_a must be labeled $\oplus$. Since T_1 is in normalized form, the root r_a of T'_a has label $\cup$. Since r_a has at least two children, we may fix a child x of r_a that is not an ancestor of the nodes labeled u and v. Let $w \in V(G)$ be the label of a leaf in the branch rooted at x. By looking at the label of the lowest common ancestor r_a, we conclude that w is not adjacent to u nor v in G. This gives the desired contradiction, as the decomposition $G = G''_1 \oplus \cdots \oplus G''_\ell$ implies that $\{u, w\} \notin E(G)$ only if $w \in V(G''_b)$ and that $\{v, w\} \notin E(G)$ only if $w \in V(G''_c)$, implying that $b = c$. $\square$

Given a graph G, we let T_G denote the unique cotree in normalized form associated with it, which is often called the *minimal cotree* of G. From an algorithmic point of view, it is clear that a graph may be easily obtained from any of its cotrees. This begs the question of whether the converse is also true, that is, whether there is an efficient algorithm that recognizes if an input graph G is a cograph and produces a (minimal) cotree associated with it. This question has been answered in the affirmative by Corneil, Perl and Stewart [13], who designed a linear time algorithm to solve this problem. By the above discussion, this linear time recognition algorithm allows one to find maximum cliques and maximum independent sets in cographs in linear time.

Another notion that may be used to characterize cographs is the following. Two vertices u and v in a graph G are *duplicates*[3] if $N(u) = N(v)$ and *coduplicates*[4] if $N[u] = N[v]$. We call u and v *siblings* if they are either duplicates or coduplicates. Siblings play an important role in the structure of cographs.

Lemma 4.2 *Two vertices v and u in a cograph are siblings if and only if they share the same parent node x in the minimal cotree. Moreover, if $x = \cup$, they are duplicates. If $x = \oplus$, they are coduplicates.*

Proof Let u and v be vertices in a cograph G, and let T_G be its minimal cotree. If the leaves corresponding to u and v have the same parent, then, given any other vertex w in G, the lowest common ancestor of the leaves corresponding to w and u in T_G is also the lowest common ancestor of the leaves corresponding to w and v.

---

[3] Also called *false twins* or *weak siblings*.

[4] Also called *true twins* or *strong siblings*.

By Lemma 4.1, u and w are adjacent if and only if v and w are adjacent. This makes u and v siblings. They are coduplicates if $x = \oplus$ and duplicates if $x = \cup$.

For the converse, assume that the leaves corresponding to u and v have distinct parent nodes x and y in T_G, respectively, and let z be their lowest common ancestor. First assume that x and z are labeled differently, and fix a vertex $w \in V(G)$ such that x is the lowest common ancestor of the leaves corresponding to u and w. Note that w exists because all leaves of T_G correspond to vertices of G and any internal node has at least two children by the definition of minimal cotree. The adjacency between u and w is determined by the label of x, and the adjacency between v and w is determined by the label of z. Thus u and v are not siblings. Of course, we reach the same conclusion if y and z are labeled differently. Next assume that x, y, and z have the same label. By the definition of minimal cotree, there is an internal node z' with the other label on the path between x and z (or between y and z, if $x = z$) in T_G. We now fix a vertex w such that z' is the lowest common ancestor of u and w. Note that the lowest common ancestor of v and w is z, so that w is a witness that u and v are not siblings, as in the previous case. $\square$

We may use this to prove that cographs are precisely the graphs that may be constructed from a singleton by a sequence of duplications and coduplications. This construction was very important for the results in [30], for instance.

Theorem 4.4 *An n-vertex graph G is a cograph if and only if there is a graph sequence $G_1, G_2, \ldots, G_n$ such that G_1 has order 1, $G_n = G$, and, for $i \in \{1, \ldots, n-1\}$, G_{i+1} is obtained from G_i by the addition of a sibling of a vertex $u \in V(G_i)$.*

Proof We first prove, by induction on n, that any cograph of order n has a sequence as in the statement. This is obvious for $n \in \{1, 2\}$, and we assume that it holds for all cographs with $n \geq 2$ vertices. Let G be a cograph with $n + 1$ vertices, and let T_G be the minimal cotree associated with it. It is clear that the node farthest from the root is a leaf corresponding to a vertex u of G, and since no internal vertex has a single child, the parent x of u has a second child, a leaf that corresponds to a vertex v. By Lemma 4.2, u and v are siblings. By the induction hypothesis, the cograph $H = G - v$ admits a sequence $G_1, G_2, \ldots, G_{n-1}$ as in the statement. In particular, $G_{n-1} = H$ and G is precisely the graph G_n obtained from H by duplicating u (if x is labeled $\cup$) or coduplicating u (if x is labeled $\oplus$), and by calling the new vertex v.

Next assume that G is an n-vertex graph that admits a sequence $G_1, G_2, \ldots, G_n$ as in the statement, and assume for a contradiction that G is not a cograph. By Theorem 4.2, G_1, G_2, and G_3 are cographs; let $i \geq 4$ be the least index such that G_i is not a cograph. Let u be the vertex of G_{i-1}, whose sibling v has been added to produce G_i. Denote $v_1, \ldots, v_4$ the vertices of G_i that induce a copy of P_4 (the labeling respects the order of the vertices along the path). Clearly, v must lie on the path; otherwise, G_{i-1} would contain an induced P_4, contradicting the minimality of i. On the other hand, since u and v are siblings, $G_i - u$ and $G_{i-1} = G_i - v$ are isomorphic. This means that any induced copy of P_4 in G_i that contains v but does not contain u would correspond to an induced copy of P_4 that contains u and does not contain v and therefore would be entirely in G_{i-1}, a contradiction. As a consequence, u and v must be both on the path. By symmetry, there are four cases: $(u, v) = (v_1, v_2)$, $(u, v) = (v_1, v_4)$, $(u, v) = (v_1, v_3)$, and $(u, v) = (v_2, v_3)$. In the first two cases, v_1 would be adjacent to v_3, a contradiction. In the third case, v_1 would be adjacent to v_4, and in the fourth case, v_2 would be adjacent to v_4, also contradictions. This concludes the proof.   □

Theorem 4.4 implies yet another characterization of cographs.

Corollary 4.2 *A graph G is a cograph if and only if every induced subgraph F of G with at least two vertices has siblings.*

Proof If G is a singleton, the result is trivially true, so consider graphs with $n \geq 2$ vertices. If G is a cograph and F is an induced subgraph of order $m \geq 2$, then F is a cograph and has siblings by Theorem 4.4. Conversely, assume that G is an n-vertex graph such that every induced subgraph with at least two vertices has siblings. Applying this property to G itself, we find siblings $u_n, v_n \in V(G)$, so that we may set $G_n = G$ and $G_{n-1} = G - v_n$. If G_{n-1} has a single vertex, we are done. Otherwise, it has at least two vertices, so that it contains siblings u_{n-1}, v_{n-1}, and we extend the sequence with $G_{n-2} = G_{n-1} - v_{n-1}$. The result follows by induction.   □

In Chaps. 5 and 7, we will describe linear time eigenvalue location algorithms for cographs based on such structural decompositions. We shall also see that, based on this, it is easy to obtain an eigenvalue-based characterization of cographs.

4.3   Tree Decomposition

A tree decomposition of a graph G may be viewed as a tree associated with G that is a witness for the fact that G may be stored efficiently. As we shall see, this decomposition is very useful for algorithmic purposes. To get some intuition, assume that we have access to an oracle that, given a graph G and any pair $\{u, v\} \in V$, tells us whether u and v are adjacent or not. Each time we ask the oracle about an adjacency, we say that we performed a *query*. Clearly, if we are told that G is an n-vertex graph and we wish to identify G, we need to perform $\binom{n}{2} = O(n^2)$ queries, as all pairs $\{u, v\} \in V$ must be investigated. However, if we also know that G has a tree decomposition of *width* k, where the width measures the complexity of the decomposition, we will see that the number of necessary queries is reduced to $O(nk)$.

The modern definition of tree decomposition was introduced in the seminal work of Robertson and Seymour [33, 34], playing a crucial role in the proof of the Graph Minor Theorem. This is a very deep result whose original proof was presented in a series of more than 20 papers, taking well over 500 pages. Despite the importance of this body of work in popularizing tree decompositions, it turns out that notions that are similar or even equivalent have been independently studied in other contexts. Some of the early work in this direction appears in [2, 24, 35, 37].

The fact that this structural description has been rediscovered many times is not surprising given its extraordinary potential for the design of efficient algorithms. Quoting Bodlaender [5]:

It appears that many problems that are intractable (e.g., NP-hard) for general graphs become polynomial or linear-time-solvable, when restricted to graphs of bounded treewidth.

In light of this, the study of structural aspects of tree decompositions and their algorithmic implications became a rich and fruitful research topic, and in a compact book, we can only skim over its surface. To dive into this topic, interested readers are referred to influential survey papers by Bodlaender [3, 5, 6].

A *tree decomposition* of a graph $G = (V, E)$ is a tree $\mathcal{T}$ with nodes $1, \ldots, m$, where each node i is associated with a set $B_i \subseteq V$, called a *bag*, satisfying the following properties:

(1) $\bigcup_{i=1}^m B_i = V$.
(2) For every edge $\{v, w\} \in E$, there exists B_i containing v and w.
(3) For any $v \in V$, the subgraph of $\mathcal{T}$ induced by the nodes that contain v is connected.

The *width* of the tree decomposition $\mathcal{T}$ is defined as $\max\{|B_i| - 1 : i \in V\}$, and the *treewidth* $tw(G)$ of graph G is the smallest k such that G has a tree decomposition of width k. In the particular case where the tree $\mathcal{T}$ is a path, the decomposition is known as a *path decomposition* of G [34]. This leads to an analogous notion of width, the *pathwidth* $pw(G)$ of graph G. Figure 4.4 depicts a tree decomposition of a graph G. In general, we observe that the distinction between the nodes of the tree and the bags associated with them is important, as different nodes may be associated with equal bags.

For any graph G, the tree $\mathcal{T}$ with a single node 1 associated with the bag $B_1 = V(G)$ is a tree decomposition of G, so that $tw(G) \leq pw(G) \leq |V| - 1$. This illustrates an important difference between a tree decomposition of a graph and other decompositions, such as cotrees of cographs and k-expressions of graphs: tree decompositions do not determine the graph, i.e., different graphs may have the same tree decomposition. As it turns out, if $G = (V, E)$ is a graph with $|V| \geq 2$, then $tw(G) = 1$ if and only if G is a forest, and $tw(G) = |V| - 1$ if and only if G is a

Fig. 4.4 A graph and a tree decomposition of width 2 associated with it. The vertices in each bag are listed within each node

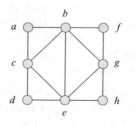
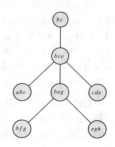

complete graph. This will be justified as part of a more general result. Before doing this, we present some easy consequences of the definition of treewidth.

Proposition 4.1 *Let $G = (V, E)$ be a graph.*

(a) If H is a subgraph of G, then $tw(H) \leq tw(G)$.
(b) If G is a graph with components $G_1, \ldots, G_\ell$, then

$$tw(G) = \max\{tw(G_1), \ldots, tw(G_\ell)\}.$$

(c) If $K \subseteq V$ induces a clique in G and $\mathcal{T}$ is a tree decomposition of G, then there exists a node i such that $K \subseteq B_i$.
(d) For all $e \in E$, the graph G/e obtained by contracting e satisfies $tw(G/e) \leq tw(G)$.

Proof Parts (a), (b), and (d) are easy. For (a), take a tree decomposition of G with minimum width and simply delete all vertices in $V(G) \setminus V(H)$ from the bags. For (b), note that part (a) implies $\max\{tw(G_1), \ldots, tw(G_\ell)\} \leq tw(G)$. For the other direction, we produce a tree decomposition of G by taking (vertex-disjoint) trees of width $tw(G_i)$ for each G_i and connecting an arbitrary node of each tree to a new node with an empty bag. For (d), any tree decomposition of G becomes a tree decomposition of G/e, where $e = \{x, y\}$, just by modifying the bags: replace each occurrence of x and y in a bag by an occurrence of the new vertex v_e. Note that the sizes of the bags remain the same except for the bags that originally contained both x and y, whose size decreases by one.

For part (c), consider a tree decomposition $\mathcal{T}$ and view it as a rooted tree by fixing an arbitrary root r. For each $v \in K$, let $i_v \in V(\mathcal{T})$ be the node that achieves the minimum

$$d(i_v, r) = \min\{d(i, r) : i \in V(\mathcal{T}), v \in B_i\}.$$

By (3) in the definition of tree decomposition, given v, there is a unique choice for i_v. Define $d = \max\{d(i_v, r) : v \in K\}$ and fix $w \in K$ such that $d(i_w, r) = d$.

We claim that $v \in B_{i_w}$ for all $v \in K$. Note that this is immediate if $d = 0$, so assume that $d > 0$. By (2) in the definition of tree decomposition, for each $v \in K$, there is a node j such that $v, w \in B_j$. Since i_w is the node closest to the root among all nodes whose bags contain w, the distance between j and r is at least d. Moreover, v must lie in the bags of all nodes on the path between j and i_v in $\mathcal{T}$. One of these nodes, call it j', is at distance exactly d from r. If $j' = i_w$, we get $v \in B_{i_w}$, as required. Otherwise, j' and i_w would be distinct nodes whose distance to the root is the same, and the path connecting them in $\mathcal{T}$ would include the root r. As a consequence, the path between j and w contains r. Since $w \in B_j \cap B_{i_w}$, this implies that $w \in B_r$, a contradiction to $d > 0$.   □

Let $G = (V, E)$ be a graph, and let K induce a maximum clique in G, that is, a clique of size $\omega(G)$. Proposition 4.1(c) tells us that any tree decomposition $\mathcal{T}$ of G contains a bag B_i such that $|B_i| \geq |K| = \omega(G)$. This leads to $tw(G) \geq \omega(G) - 1$.

More generally, combining this with Proposition 4.1(a) and (d), we see that if G can be turned into a clique K_k by a series of edge contractions and vertex deletions, then $tw(G) \geq k - 1$. A structural decomposition of Gavril [23] implies that this inequality is tight for all *chordal graphs*. A graph $G = (V, E)$ is said to be *chordal* if any cycle in it that consists of four or more vertices has a chord, that is, there is an edge in the graph that is not part of the cycle connecting two vertices of the cycle. In other words, a graph is chordal if it is $\mathcal{F}$-free for $\mathcal{F} = \{C_\ell : \ell \geq 4\}$. Note that forests and complete graphs are chordal, leading to $tw(G) = 1$ if G is a forest and $tw(G) = |V(G)| - 1$ if G is a complete graph.

4.4   Nice Tree Decomposition

In this book, it will be particularly convenient to consider tree decompositions with a particular structure, which have been introduced by Kloks [27] and are called *nice tree decompositions*. Given a graph G, this is a rooted tree decomposition $\mathcal{T}$ for which the root has an empty bag and each node i is of one of the following types:

(a) **(Leaf)**. The node i is a *leaf* of $\mathcal{T}$.
(b) **(Introduce)**. The node i *introduces* vertex v, that is, it has a single child j, $v \notin B_j$ and $B_i = B_j \cup \{v\}$.
(c) **(Forget)**. The node i *forgets* vertex v, that is, i has a single child j, $v \notin B_i$ and $B_j = B_i \cup \{v\}$.
(d) **(Join)**. The node i is a *join*, that is, it has two children j and ℓ, where $B_i = B_j = B_\ell$.

Before describing the advantages of the additional structure given by a nice tree decomposition, we observe that some authors do not require the root's bag to be empty. However, any tree decomposition $\mathcal{T}$ that satisfies (a)–(d), but does not satisfy this additional requirement, may be turned into a nice tree decomposition by appending a path of length at most $k + 1$ to the root of $\mathcal{T}$, where all nodes have type Forget. The length of the path is due to the width of $\mathcal{T}$. Because of property (3) in the definition of tree decomposition, we see that every vertex of G can be forgotten at most once in $\mathcal{T}$. Given that the root has an empty bag (and property (1) in the definition of tree decomposition), we conclude that each vertex of G must be forgotten exactly once in a nice tree decomposition.

Figure 4.5 depicts a nice tree decomposition of the graph of Fig. 4.4.

Now, to illustrate the usefulness of the decomposition, we go back to the situation described at the beginning of Sect. 4.3, where we wished to identify a graph G of order n by querying possible adjacencies. Since, for any edge $\{u, v\}$ of G, there is a bag in any tree decomposition that contains both u and v, the following occurs for any nice tree decomposition: either u is in the bag that forgets v or vice versa. This means that, to capture all the adjacencies of G, it suffices that, for each node ℓ of type Forget, we ask the oracle about pairs $\{u, v\}$ such that u is the vertex forgotten

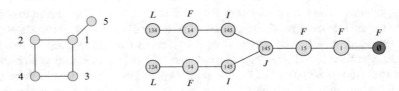

Fig. 4.5 A graph and a nice tree decomposition associated with it. The root of the tree is on the right, and the nodes are of type L (leaf), I (introduce), F (forget), or join (J)

at node ℓ and $v \in B_\ell$. A more formal version of this argument appears in the proof of Theorem 4.5 below. In the example of Fig. 4.5, this strategy would require seven queries to identify the five edges of the graph, instead of the 10 queries that would be needed without the nice tree decomposition.

Interestingly, despite the additional structure, a nice tree decomposition may be efficiently obtained from an arbitrary tree decomposition. More precisely, one may show that, if G is a graph of order n and we are given an arbitrary tree decomposition of G with width k and m nodes, it is possible to turn it into a nice tree decomposition of G with $O(n)$ nodes and width at most k in time $O((k \log k)m + kn)$. For future reference, we state a more precise version as a lemma.

Lemma 4.3 *Let G be a graph of order n with a tree decomposition $\mathcal{T}$ of width k with m nodes. Then G has a nice tree decomposition of the same width k with fewer than $4n - 1$ nodes can be computed from $\mathcal{T}$ in time $O((k \log k)m + kn)$.*

Proof Let $\mathcal{T}$ be a tree decomposition of G with width k with m nodes, and fix an arbitrary node i as root. At the beginning, we sort the vertices in each bag in increasing order (according to some arbitrary pre-determined order). This may be done in time $O((k \log k)m)$ using a standard sorting algorithm such as MergeSort, for instance. We modify the tree decomposition in a sequence of depth-first traversals. In the first traversal, every node whose bag is contained in the bag of its parent is merged with the parent, that is, the node is removed and its children are connected to the parent.

In the second traversal, whenever the bag of a node has size less than the size of the bag of its parent, we add some vertices from the parent's bag until both bags have the same size. At this point, the bags of children are at least as large as the bags of their parents, and they are never contained in the bag of their parent. In the third traversal, each node i with $c \geq 2$ children and bag B_i is replaced by a binary tree with exactly c leaves whose nodes are all assigned the bag B_i. Each child of i in the original tree becomes the single child of one of the leaves of this binary tree. At this point, all nodes have at most two children, and those with two children are Join nodes. In the fourth traversal, for any node i with a single child j, if necessary, replace the edge $\{i, j\}$ by a path such that the bags of the nodes along the path differ by exactly one vertex in each step. This is done from j to i by a sequence of nodes, starting with a Forget node, then alternating between Introduce and Forget nodes, and possibly ending with a sequence of Forget nodes in case the child's bag is larger

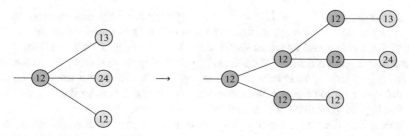

Fig. 4.6 The transformation performed in the third traversal, where a vertex is replaced by a binary tree. The root of the decomposition is located toward the left

Fig. 4.7 The transformations performed in the fourth traversal, which creates an alternating path of Forget and Introduce nodes. The root of the decompositions is located toward the left

than its parent's bag. To ensure that property (3) of a tree decomposition is satisfied, each vertex in the symmetric difference of the original bags B_i and B_j produces a single Forget or Introduce node. The operations performed in the third and fourth traversals are illustrated in Figs. 4.6 and 4.7.

To finish our construction, if the root i has a nonempty bag B_i, we append a path to the root where each node is a forget node, until we get an empty bag, which becomes the new root. Note that at most $k + 1$ nodes are appended to the path, as $|B_i| \leq k + 1$. We call this nice tree decomposition $\mathcal{T}'$. We claim that $\mathcal{T}'$ is a nice tree decomposition of G with the same width as $\mathcal{T}$. It is easy to see that properties (1) and (3) in the definition of tree decomposition are not violated after each traversal and that all nodes in $\mathcal{T}$ have one of the types in the definition of nice tree decomposition. Moreover, bag sizes have only been increased in the second traversal, but the size of a new bag was always bounded by the size of a bag already in the tree. Finally, we see that $\mathcal{T}'$ is a decomposition of G; in the sense that property (2) in the definition of tree decomposition is satisfied. To this end, if two vertices $u, v \in V(G)$ lie a bag of $\mathcal{T}$, note that one of the bags of the new tree must contain the largest bag originally containing u and v, and therefore, it is a tree decomposition of G.

Next, we count the number m' of nodes of $\mathcal{T}'$. The Leaf nodes have degree 1, the Join nodes have degree 3 (unless one of the Join nodes is the root and has degree 2), and Forget and Introduce nodes have degree 2 (unless one of the Forget nodes is the root and has degree 1). Let m'_F, m'_I, m'_J, and m'_L denote the number of Forget, Introduce, Join, and Leaf nodes in $\mathcal{T}'$, respectively. Since the sum of degrees is $2m' - 2$, we have

$$2(m'_F + m'_I + m'_J + m'_L) - 2 = 3m'_J + 2(m'_I + m'_F) + m'_L - 1,$$

which leads to $m'_J = m'_L - 1$. Recall that every vertex is forgotten exactly once, so $m'_F = n$. To see that $m'_L \leq n$, first note that $m'_L = 1$ if there are no Join nodes. Otherwise, the first and fourth traversals ensure that there is at least one Forget node on the path between each Leaf node and the first Join node on its path to the root, so that $m'_L \leq m'_F \leq n$. It follows that $m'_J \leq n - 1$. The fourth traversal ensures that the single child of every Introduce node is a Forget node, so that we also have $m'_I \leq m'_F \leq n$. Therefore, $\mathcal{T}'$ has at most $4n - 1$ nodes.

To conclude the proof, we compute the time required for this transformation. In the first traversal, for each node of $\mathcal{T}$, we need to check whether its bag is contained in its parent's bag. This is done in time $O(km)$ since we are assuming that the bags are sorted. Note that the other traversals cannot decrease the number of nodes in the tree, so the trees traversed in the remaining steps have size $O(n)$ by the above discussion. The second traversal also requires comparing the bag of each node with its parent's bag, so it takes time $O(kn)$. The third traversal requires to verify the degree of each node in the tree and possibly create a number of new nodes with bags of size at most $k + 1$. This takes time $O(kn)$ because we know that the number of nodes is at most $4n$ *after* all the new nodes have been created. The final traversal also takes order $O(kn)$ because we only need to verify node degrees, compute the symmetric difference of the content of bags of size at most $k + 1$, and create new nodes with bags of size at most $k+1$. We again observe that the number of nodes is at most $4n$ *after* all the new nodes have been created. Overall, the changes performed starting from the first traversal take time $O(km + kn)$. The first term is bounded above by $O((k \log k)m)$, which is a worst case estimate to sort the elements in the bags of the original decomposition, assuming that they are initially unsorted.   □

In Chap. 6, nice decompositions will be the basis of a dynamic programming algorithm to find a diagonal matrix that is congruent to an input symmetric matrix M. Here, we use it to provide a short proof of the fact that graphs with bounded treewidth must be sparse.

Theorem 4.5 *Let $k \geq 1$, and consider a graph $G = (V, E)$ with $|V| \geq k + 1$. If G has treewidth at most k, then $|E| \leq k|V| - \binom{k+1}{2}$.*

Proof We fix $k \geq 1$ and use induction on $n = |V(G)|$. The base case is $n = k + 1$, where the results are trivial because $|E| \leq \binom{k+1}{2} = k(k + 1) - \binom{k+1}{2}$. Assume that the result holds for all graphs with $n \geq k + 1$ vertices and treewidth at most k, and let G be a graph with $n + 1$ vertices such that $tw(G) \leq k$. Fix a nice tree decomposition $\mathcal{T}$ of G whose width is at most k, and call its root r. We may assume that the root has an empty bag. Consider a forget node $i \in V(\mathcal{T})$ whose distance to r is maximum, let j be its single child, and let v be the vertex that has been forgotten. Since $v \notin B_i$, $v \in B_j$, and the subgraph of $\mathcal{T}$ induced by the nodes whose bags contain v is connected, we conclude that v can only appear in the bag of j and its descendants. Moreover, all possible neighbors of v in G lie in B_j, as there are no forget nodes in the subtree rooted at j by our choice of i. Hence, the degree of v in G is at most $|B_j| - 1 \leq k$.

As $G - v$ is an n-vertex subgraph of G, it has treewidth at most k by Proposition 4.1(a). By induction,

$$|E(G)| \leq |E(G - v)| + k \leq \left(kn - \binom{k+1}{2} \right) + k = k(n+1) - \binom{k+1}{2},$$

as required. $\qquad\qquad\qquad\qquad\qquad\qquad\qquad\qquad\qquad\qquad\qquad\qquad\qquad\qquad\square$

As we already mentioned, having a tree decomposition of a graph $G = (V, E)$ with small width is quite useful algorithmically. Like other graph decompositions, the treewidth has been used to design algorithms for NP-hard or even harder problems that are efficient on graphs of bounded width. In complexity theory, this means that the problems are *fixed parameter tractable* (FPT), namely they admit an algorithm with running time $O(f(k)n^c)$, where c is a constant and f is an arbitrary computable function that depends on a parameter k (here the width of the graph). Even if f is exponential, for small values of k, such algorithms are often very practical. We refer to Niedermeier [31] for a general introduction to fixed parameter tractability and to Bodlaender and Koster [7] for algorithms of this type that concern specifically the treewidth. This is part of the more general area of parameterized complexity [18, 19].

Unfortunately, computing the treewidth of a graph G is NP-complete in general [1]; however, an important result of Bodlaender [4] shows that, for any fixed constant k, there is an algorithm with running time $f(k)n$ to decide whether an n-vertex graph has treewidth at most k and outputs a tree decomposition if the answer is positive (where $f(k) = k^{O(k^3)}$). Moreover, there are polynomial-time algorithms to obtain decompositions for many graph classes, including cographs and distance-hereditary graphs.

In addition to this, a lot of effort has been put in finding good approximation algorithms. For instance, Bodlaender et al. [8] devised an algorithm that, given any n-vertex graph and any integer k, runs in time $2^{O(k)}n$ and either outputs that the treewidth of G is larger than k, or gives a tree decomposition of G of width at most $5k + 4$. More recently, Fomin et al. [21] described an algorithm that has running time $O(k^7 n \log n)$ and either correctly reports that the treewidth of G is larger than k, or constructs a tree decomposition of G of width $O(k^2)$. We should mention that it is a longstanding open question whether there is a constant factor approximation algorithm for the treewidth in general graphs.[5]

---

[5] A positive answer is a polynomial-time algorithm that, for a fixed positive constant C and any input graph G, outputs a tree decomposition $\mathcal{T}$ whose width is at most $C \cdot tw(G)$.

4.5 Clique Decomposition

In Sect. 4.3, we introduced a structural description of graphs known as tree decomposition, and we observed that it is quite useful for algorithmic purposes, particularly when the treewidth is bounded by a constant. On the other hand, we have also shown that n-vertex graphs whose treewidth is bounded by k must have fewer than nk edges, so that the impact of tree decompositions is greater for sparse graphs. It turns out that even well-behaved graph classes, such as the class of cographs, contain graphs with unbounded treewidth. In fact, a very recent result of Lozin and Razgon [29] shows that all elements in the hereditary family of $\mathcal{F}$-free graphs have bounded treewidth if and only if $\mathcal{F}$ contains graphs of the following four types: a complete graph, a complete bipartite graph, a forest in which every connected component has at most three leaves (known as a *tripod*), and the line graph of a tripod.

A new decomposition, called *clique decomposition*, was introduced in 2000 by Courcelle and Olariu [16] with the aim of capturing the structure of denser graphs with a "bounded" description and has also proven to be instrumental in the design of algorithms. In their seminal paper, Courcelle and Olariu wrote:

> Many NP-complete problems have linear complexity on graphs with tree-decompositions of bounded width. We investigate alternate hierarchical decompositions that apply to wider classes of graphs and still enjoy good algorithmic properties. These decompositions are motivated and inspired by the study of vertex-replacement context-free graph grammars. The complexity measure of graphs associated with these decompositions is called *clique-width*.

For more information about context-free graph grammars, we refer to [15, 17] and the references therein.

The above description by Courcelle and Olariu hints that, unlike tree decompositions, the clique-width is based on a logical algebraic description of graphs. Given a positive integer k, a *k-expression* is an algebraic expression formed from atoms $i(v)$, two unary operations $\eta_{i,j}$ and $\rho_{i \to j}$, and a binary operation $\oplus$, which produce a labeled graph with labels in the set $[k] = \{1, \ldots, k\}$[6] as follows:

(a) $i(v)$ creates a vertex v with label i, where $i \in [k]$.
(b) $\eta_{i,j}$ creates edges (not already present) between every vertex with label i and every vertex with label j for $i \neq j$.
(c) $\rho_{i \to j}$ changes all labels i to j.
(d) $\oplus$ produces the disjoint union of two labeled graphs.

Finally, the graph generated by a k-expression[7] is obtained by deleting the labels. The *clique-width* $cw(G)$ of a graph G is the smallest k such that the graph can be

---

[6] In this context, the *labels* are values assigned to the vertices, not the *names* of the vertices. In particular, several vertices may have the same label.

[7] For simplicity, we call the expression a k-expression even if some labels in $[k]$ are left out, that is, a k-expression is also a k'-expression for every $k' \geq k$.

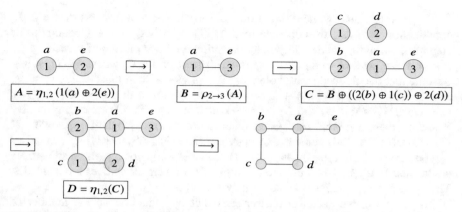

Fig. 4.8 Generating the graph G associated with the 3-expression in (4.1)

constructed by a k-expression, that is, the least number of labels such that there is an expression for G with these labels. It follows immediately from the definition that a graph G satisfies $cw(G) = 1$ if and only if it is edgeless. Note that any n-vertex graph G may be trivially generated by an n-expression where each vertex v is created with its own label $\phi(v)$, and we create each edge $\{u, v\} \in E(G)$ by applying the operation $\eta_{\phi(u),\phi(v)}$. The challenge is to find a k-expression for G with small k. Figure 4.8 depicts how the 3-expression

$$\eta_{1,2}\left(\rho_{2\to3}\left(\eta_{1,2}\left(1(a) \oplus 2(e)\right)\right) \oplus \left((2(b) \oplus 1(c)) \oplus 2(d)\right)\right) \qquad (4.1)$$

generates a graph G. Since G is not a cograph, as $G - d$ is isomorphic to P_4, Proposition 4.2 below implies that $cw(G) = 3$.

It turns out that many dense graphs have small clique-width. For instance, all cliques have clique-width 2. Regarding graph classes mentioned in this chapter, we have the following.

Proposition 4.2 *The following hold for a graph G.*

(a) If H is an induced subgraph of G, then $cw(H) \leq cw(G)$.
(b) G is a cograph if and only if $cw(G) \leq 2$.
(c) If G is a tree, then $cw(G) \leq 3$.

Moreover, $cw(G) = 3$ in (c) if and only if G contains an induced copy of P_4.

Proof To prove (a), start with a k-expression for G, where $k = cw(G)$, and remove all references to vertices in $V(G) \setminus V(H)$, that is, remove the atoms $i(v)$ for all $v \in V(G) \setminus V(H)$. Until an algebraically valid expression is obtained, repeatedly delete operations $\oplus$ with fewer than two nonempty arguments, and operations of type $\eta_{i,j}$ or $\rho_{i\to j}$ applied to an empty graph. The resulting expression is a k-expression that generates H.

To prove (b), an argument based on case analysis shows that $cw(P_4) = 3$. In combination with (a), we conclude that $cw(G) \geq 3$ if G is not a cograph. For cographs, we show that $cw(G) \leq 2$ by induction on the number n of vertices of G. If $n = 1$, we have $G = K_1$ and $1(v)$ is a 1-expression for G, where $V(G) = \{v\}$. Assume that the claimed result holds for all cographs with at most $n \geq 1$ vertices, and let G be a cograph on $n + 1 > 1$ vertices. We know that $G = G_1 \cup G_2$ or $G = G_1 \oplus G_2$, where G_1 and G_2 are vertex-disjoint cographs of order at most n. By induction, these two graphs are generated by 2-expressions s_1 and s_2, respectively. If $G = G_1 \cup G_2$, then G is generated by $s_1 \oplus s_2$. If $G = G_1 \oplus G_2$, then it is generated by $\eta_{1,2}(\rho_{2\to1}(s_1) \oplus \rho_{1\to2}(s_2))$, as $\rho_{2\to1}(s_1)$ produces a copy of G_1 where all vertices are labeled 1 and $\rho_{1\to2}(s_2)$ produces a copy of G_2 where all vertices are labeled 2. The operation $\eta_{1,2}$ produces all edges between G_1 and G_2.

To prove (c), we establish a stronger result by induction on the number of vertices, namely that for any tree T, any fixed $v \in V(T)$, and any fixed $i \in [3]$, there is a 3-expression that generates a labeled copy of T for which v is the single vertex with label i. For $T = K_1$ with $V(T) = \{v\}$, $i(v)$ is the required expression.

Assume that the result holds for all trees with at most $n \geq 1$ vertices. Let T be a tree on $n + 1$ vertices, and fix $v \in V(T)$ and $i \in [3]$. Without loss of generality, assume that $i = 1$. Let $T_1, \ldots, T_\ell$ be the components of $T - v$, where $\ell \geq 1$. Let v_j be the neighbor of v in T_j. By induction, for each $j \in [\ell]$, we find a 3-expression s_j generating T_j for which v_j is the only vertex with label 2. The required expression for T is given by

$$s = \eta_{1,2}(1(v) \oplus \rho_{1\to3}(s_1) \oplus \cdots \oplus \rho_{1\to3}(s_\ell)),$$

where we omit parentheses between successive applications of the binary operation $\oplus$ because it is associative. This concludes the proof.   □

A survey about the clique-width of several graph classes has been published by Kamiński, Lozin and Milanič [26]. As was the case for the treewidth, computing the clique-width is NP-complete for arbitrary graphs [20]. However, unlike what happens for the treewidth, no polynomial algorithm to recognize whether a graph has bounded clique-width k is known for any fixed $k > 3$; there are algorithms for $k \leq 3$ [14]. For fixed k, Oum and Seymour [32] devised an $O(n^9 \log n)$-time approximation algorithm that either gives a $(2^{3k+2} - 1)$-expression for an input n-vertex graph G, or provides a witness that G does not have clique-width $k + 1$. Thus, algorithms based on the clique-width typically assume that a k-expression of the graph is part of the input.

Interestingly, graphs with bounded treewidth have bounded clique-width, albeit with exponential dependency. To be precise, Corneil and Rotics [11] proved that a graph with treewidth k has clique-width at most $3 \cdot 2^{k-1}$ and that, for any k, there is a graph with treewidth k and clique-width at least $2^{\lfloor k/2 \rfloor - 1}$.

Since edges can be added at any time, a subexpression of a k-expression does not necessarily generate an induced subgraph of the graph generated by the entire expression. This is a drawback for the design of a dynamic programming diagonalization algorithm that uses the expression. This motivates the definition of a closely related expression, which is the subject of the next section.

4.6 Slick Clique Decomposition

A *slick expression* is a hierarchical decomposition that is influenced by clique expressions but has the additional property that subexpressions define induced subgraphs. The associated graph parameter is the *slick clique-width*. In this new definition, a single operator is used for performing the union, creating edges, and relabeling. A *slick expression* is an expression formed from atoms $i(v)$ and a binary operation $\oplus_{S,L,R}$, where L, R are functions from $[k]$ to $[k]$ and S is a binary relation on $[k]$, which produce a graph as follows:

(a) $i(v)$ creates a vertex v with label i, where $i \in [k]$.
(b) Given two graphs G and H whose vertices have labels in $[k]$, the labeled graph $G \oplus_{S,L,R} H$ is obtained as follows. Starting with the disjoint union of G and H, add edges from every vertex labeled i in G to every vertex labeled j in H for all $(i, j) \in S$. Afterward, every label i of the *left component* G is replaced by $L(i)$, and every label j of the *right component* H is replaced by $R(j)$.

The graph constructed by a slick expression is obtained by deleting the labels of the labeled graph produced by it. The *slick clique-width* of a graph G, denoted $scw(G)$, is the smallest k such that the graph can be constructed by a slick expression. For instance, the graph G in Fig. 4.8 may be generated by the slick expression below, where $S = \{(1, 2), (2, 1)\}$, *id* denotes the identity map, and $2 \to 3$ denotes the map that relabels 2 with 3, leaving other labels unchanged:

$$\left(\left(1(b) \oplus_{\{(1,2)\},id,id} 2(c)\right) \oplus_{S,id,2\to3} \left(2(a) \oplus_{\{(2,1)\},id,id} 1(d)\right)\right) \oplus_{\{(3,1)\},id,id} 1(e).$$

$$(4.2)$$

In fact, the slick expression in (4.2) is not as economical as possible, and the following slick expression generates the same graph G:

$$\left(\left(1(b) \oplus_{\{(1,2)\},id,id} 2(c)\right) \oplus_{S,2\to1,id} \left(2(a) \oplus_{\{(2,1)\},id,id} 1(d)\right)\right) \oplus_{\{(2,1)\},id,id} 1(e).$$

$$(4.3)$$

Since a single operator performs the join, edge creation, and relabeling, a slick expression may be associated with a tree in a way that closely resembles the relation between cographs and cotrees. As shown in Fig. 4.9, a slick expression for a graph G of order n may be represented as a binary parse tree T having $2n - 1$ nodes, where the n leaves are labeled by the operators $i(v)$ and the internal nodes contain operations of type $\oplus_{S,L,R}$. Additionally, the left child corresponds to the root of the left component, while the right child corresponds to the root of the right component.

Fig. 4.9 A parse tree for the
slick expression (4.3)

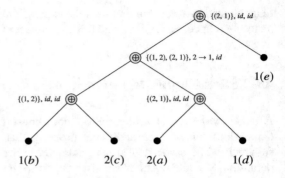

The following proposition records properties of this binary parse tree that will be crucial for our algorithm in Chap. 7. They are straightforward consequences of the definition of slick decomposition.

Proposition 4.3 *Let $G = (V, E)$ be a graph generated by a slick expression s, and let T be the binary parse tree associated with this expression.*

(a) *Two vertices $u, v \in V$ are adjacent in G if and only if the operation at the lowest common ancestor of the leaves corresponding to them in the tree places an edge between them.*

(b) *Let s' be the subexpression of s associated with a subtree T' of T rooted at one of its nodes. The graph H generated by s' is the subgraph of G induced by the vertices corresponding to the leaves of the subtree.*

We note that property (b) in the above proposition is also satisfied by another variation of clique-width, known as NLC width, due to Wanke [36]. Both NLC expressions and slick expressions produce edges between the left and right components using a relation S, but slick expressions are allowed to use different relabeling functions for the left and right components, while NLC expressions perform relabeling using the unary operations $\rho_{i \rightarrow j}$, as in the usual k-expressions.

We conclude this section with properties of slick clique decompositions. We start with two basic facts.

Proposition 4.4 *The following statements hold for a graph $G = (V, E)$.*

(a) *If H is an induced subgraph of G, then $scw(H) \le scw(G)$.*

(b) *$scw(G) = 1$ if and only if G is a cograph.*

Proof In part (a), we may assume that G has more than one vertex. Given an induced subgraph H of G, we transform the parse tree T_G of G into a parse tree T_H for H that uses the same labels. If $H = G$, then $T_H = T_G$. Otherwise, H can be constructed by a finite sequence of vertex deletions so it suffices to consider $H = G - a$ for some vertex a. Let u be the parent node of the leaf corresponding to a in T_G, say $u = \oplus_{S,L,R}$. Without loss of generality, assume a is a right child of u. If u is the root of T_G, then T_H is the subtree rooted at the left child of u. Hence, we may assume that u has a parent v. Without loss of generality, suppose that u is a left child of v. There are two cases:

Case 1: The left child of u is a vertex b, with initial label i. Then we form T_H by first removing u and a and letting b be the new left child of v with initial label $L(i)$.

Case 2: The left child of u is a node $w = \oplus_{S',L',R'}$. To obtain T_H, we remove u and a and make w the new left child of v but redefine w as $w' = \oplus_{S',L\circ L',L\circ R'}$, using function composition.

Note that these operations delete a and its incident edges but preserve all other vertices and edges while creating a valid parse tree.

For part (b), we use the definition of cograph given by the family $\mathcal{D}$ of Theorem 4.2. Assume $scw(G) = 1$. Since there is a single label available, the functions L and R must be identity maps, so that $\oplus_{S,L,R}$ either creates a disjoint union (if $S = \emptyset$) or a join (if $S = \{(1,1)\}$, which adds all possible edges between the left and right components). An induction on the expression size shows that G is a cograph. For the converse, a simple induction on n shows that any n-vertex cograph G can be constructed with a slick expression using label 1 only, as the union and the join can be performed as described above. □

It turns out that the concepts of clique-width and slick clique-width are linearly related, emphasizing the close connection between these two parameters.

Theorem 4.6 *If G is a graph, then $scw(G) \leq cw(G) \leq 2scw(G)$.*

The proof of Theorem 4.6 [22, Theorem 1] provides a linear time (for fixed k) algorithm for translating a k-expression into an equivalent slick k-expression and for translating a slick k-expression into an equivalent $2k$-expression. Because of this connection, an efficient algorithm for one representation implies an efficient algorithm for the other.

References

1. Arnborg, S., Corneil, D.G., Proskurowski, A.: Complexity of finding embeddings in a k-tree. SIAM J. Algebraic Discrete Methods **8**(2), 277–284 (1987). https://doi.org/10.1137/0608024.
2. Bertelè, U., Brioschi, F.: Nonserial Dynamic Programming. Elsevier (1972)
3. Bodlaender, H.L.: A tourist guide through treewidth. Acta Cybernetica **11**(1–2), 1–21 (1993). https://cyber.bibl.u-szeged.hu/index.php/actcybern/article/view/3417
4. Bodlaender, H.L.: A linear-time algorithm for finding tree-decompositions of small treewidth. SIAM J. Comput. **25**(6), 1305–1317 (1996). https://doi.org/10.1137/S0097539793251219.
5. Bodlaender, H.L.: A partial k-arboretum of graphs with bounded treewidth. Theoret. Comput. Sci. **209**(1), 1–45 (1998). https://doi.org/10.1016/S0304-3975(97)00228-4. https://www.sciencedirect.com/science/article/pii/S0304397597002284
6. Bodlaender, H.L.: Treewidth of Graphs, pp. 2255–2257. Springer New York, New York, NY (2016)
7. Bodlaender, H.L., Koster, A.M.C.A.: Combinatorial optimization on graphs of bounded treewidth. Comput. J. **51**(3), 255–269 (2008)

8. Bodlaender, H.L., Drange, P., Dregi, M.S., Fomin, F.V., Lokshtanov, D., Pilipczuk, M.: A $c^k n$ 5-approximation algorithm for treewidth. SIAM J. Comput. **45**(2), 317–378 (2016). https://doi.org/10.1137/130947374.
9. Brandstädt, A., Le, V.B., Spinrad, J.P.: Graph Classes: A Survey. Soc. Ind. Appl. Math. (1999). https://doi.org/10.1137/1.9780898719796.
10. Chudnovsky, M., Robertson, N., Seymour, P.D., Thomas, R.: The strong perfect graph theorem. Ann. Math. **164**(1), 51–229 (2006)
11. Corneil, D.G., Rotics, U.: On the relationship between clique-width and treewidth. SIAM J. Comput. **34**(4), 825–847 (2005). https://doi.org/10.1137/S0097539701385351.
12. Corneil, D.G., Lerchs, H., Burlingham, L.S.: Complement reducible graphs. Discrete Appl. Math. **3**(3), 163–174 (1981). https://doi.org/10.1016/0166-218X(81)90013-5.
13. Corneil, D.G., Perl, Y., Stewart, L.K.: A linear recognition algorithm for cographs. SIAM J. Comput. **14**(4), 926–934 (1985). https://doi.org/10.1137/0214065.
14. Corneil, D.G., Habib, M., Lanlignel, J.M., Reed, B., Rotics, U.: Polynomial-time recognition of clique-width ≤ 3 graphs. Discrete Appl. Math. **160**(6), 834–865 (2012). https://doi.org/10.1016/j.dam.2011.03.020. https://www.sciencedirect.com/science/article/pii/S0166218X11001144. Fourth Workshop on Graph Classes, Optimization, and Width Parameters Bergen, Norway, October 2009
15. Courcelle, B., Engelfriet, J.: Graph Structure and Monadic Second-Order Logic: A Language-Theoretic Approach. Encyclopedia of Mathematics and Its Applications. Cambridge University Press (2012). https://doi.org/10.1017/CBO9780511977619
16. Courcelle, B., Olariu, S.: Upper bounds to the clique width of graphs. Discrete Appl. Math. **101**(1–3), 77–114 (2000)
17. Courcelle, B., Engelfriet, J., Rozenberg, G.: Handle-rewriting hypergraph grammars. J. Comput. Syst. Sci. **46**(2), 218–270 (1993)
18. Cygan, M., Fomin, F.V., Kowalik, L., Lokshtanov, D., Marx, D., Pilipczuk, M., Pilipczuk, M., Saurabh, S.: Parameterized Algorithms, 1st edn. Springer, Incorporated (2015)
19. Downey, R.G., Fellows, M.R.: Fundamentals of Parameterized Complexity, 1st edn. Springer, Incorporated (2016)
20. Fellows, M.R., Rosamond, F.A., Rotics, U., Szeider, S.: Clique-width minimization is NP-hard (extended abstract). In: STOC'06: Proceedings of the 38th Annual ACM Symposium on Theory of Computing, pp. 354–362. ACM, New York (2006). https://doi.org/10.1145/1132516.1132568.
21. Fomin, F.V., Lokshtanov, D., Saurabh, S., Pilipczuk, M., Wrochna, M.: Fully polynomial-time parameterized computations for graphs and matrices of low treewidth. ACM Trans. Algorithms **14**(3), 34:1–34:45 (2018)
22. Fürer, M., Hoppen, C., Jacobs, D.P., Trevisan, V.: Eigenvalue location in graphs of small clique-width. Linear Algebra Appl. **560**, 56–85 (2019)
23. Gavril, F.: The intersection graphs of subtrees in trees are exactly the chordal graphs. J. Combin. Theory B **16**(1), 47–56 (1974). https://doi.org/10.1016/0095-8956(74)90094-X. https://www.sciencedirect.com/science/article/pii/009589567490094X
24. Halin, R.: S-functions for graphs. J. Geometry **8**(1), 171–186 (1976)
25. Johnson, D.S.: The np-completeness column: an ongoing guide. J. Algorithms **6**(3), 434–451 (1985). https://doi.org/10.1016/0196-6774(85)90012-4. https://www.sciencedirect.com/science/article/pii/0196677485900124
26. Kamiński, M., Lozin, V.V., Milanič, M.: Recent developments on graphs of bounded clique-width. Discrete Appl. Math. **157**(12), 2747–2761 (2009). https://doi.org/10.1016/j.dam.2008.08.022
27. Kloks, T.: Treewidth: Computations and Approximations. Lecture Notes in Computer Science, vol. 842. Springer (1994)
28. Lovász, L.: Normal hypergraphs and the perfect graph conjecture. Discrete Math. **2**(3), 253–267 (1972). https://doi.org/10.1016/0012-365X(72)90006-4. https://www.sciencedirect.com/science/article/pii/0012365X72900064
29. Lozin, V., Razgon, I.: Tree-width dichotomy (2021)

30. Mohammadian, A., Trevisan, V.: Some spectral properties of cographs. Discrete Math. **339**(4), 1261–1264 (2016). https://doi.org/10.1016/j.disc.2015.11.005. https://www.sciencedirect.com/science/article/pii/S0012365X15004045

31. Niedermeier, R.: Invitation to Fixed-Parameter Algorithms. Oxford University Press (2006)

32. Oum, S., Seymour, P.: Approximating clique-width and branch-width. J. Combin. Theory B **96**(4), 514–528 (2006). https://doi.org/10.1016/j.jctb.2005.10.006. https://www.sciencedirect.com/science/article/pii/S0095895605001528

33. Robertson, N., Seymour, P.D.: Graph minors I. Excluding a forest. J. Combin. Theory B **35**, 39–61 (1983)

34. Robertson, N., Seymour, P.D.: Graph minors II. Algorithmic aspects of tree-width. J. Algorithms **7**(3), 309–322 (1986)

35. Scheffler, P.: "Die Baumweite von Graphen als ein Maß für die Kompliziertheit algorithmischer Probleme". R-MATH-4/89. PhD thesis, Akademie der Wissenschaften der DDR, Karl-Weierstraß-Institut für Mathematik Berlin (1989)

36. Wanke, E.: k-NLC graphs and polynomial algorithms. Discrete Applied Math. **54**(2), 251–266 (1994). https://doi.org/10.1016/0166-218X(94)90026-4. https://www.sciencedirect.com/science/article/pii/0166218X94900264

37. Wimer, T.: Linear algorithms in k-terminal graphs. PhD thesis, Department of Computer Science, Clemson University (1987)

Chapter 5
Locating Eigenvalues in Cographs

5.1 Diagonalizing a Row and Column

Inspired by the diagonalization algorithm for matrices associated with trees in Chap. 3, given a cograph G with adjacency matrix A, and $x \in \mathbb{R}$, we will find a diagonal matrix D congruent to the matrix $M = A + xI$. This algorithm was originally presented in [11]. By Corollary 3.1, the number of positive/negative/zero values of D is the number of eigenvalues of A greater than/smaller than/equal to $-x$, respectively.

Recall that the algorithm in Chap. 3 is iterative, where each iteration eliminates off-diagonal entries of the row and column corresponding to a vertex v by performing elementary row and column operations on it, using the rows and columns corresponding to its children. Each row operation was followed by the *same* column operation, as this maintains congruence by Eq. (2.10).

In this chapter, we will again use elementary rows and column operations to eliminate off-diagonal entries while preserving congruence. To do this, we will use the fact that cographs are described by cotrees (Theorem 4.3) and that any nontrivial cograph contains siblings (Corollary 4.2). The rows/columns of the adjacency matrix corresponding to two vertices that are siblings have the same entries, except possibly in two positions. As we explain below, this means that after performing an elementary operation, most of the nonzero elements in one of the rows/columns are removed. Since each row/column is diagonalized just once, this allows one to obtain a linear time diagonalization algorithm.

The algorithm operates bottom-up on the minimal cotree T_G of the input cograph G. At each node of the cotree, off-diagonal entries of a partially diagonalized matrix M will be annihilated. In this section, we describe a basic step of the algorithm.

Let T_G be the normalized cotree of G, and let $\{v_k, v_j\}$ be a pair of siblings. Recall from Chap. 4 that Lemma 4.2 tells us they have the same parent w. We assume the diagonal values d_k and d_j of rows (and columns) k and j, respectively, may have been modified by previous computations, but that all off-diagonal entries in these

© The Author(s), under exclusive license to Springer Nature Switzerland AG 2022
C. Hoppen et al., *Locating Eigenvalues in Graphs*, SpringerBriefs in Mathematics,
https://doi.org/10.1007/978-3-031-11698-8_5

rows and columns have the same value as in the input matrix. In particular, all off-diagonal entries are either 0 or 1. The goal is to annihilate off-diagonal 1s in the row and column corresponding to v_k, maintaining congruence to M. Let w be the parent of the siblings in T_G. There are two cases.

Case 1: $w = \oplus$

By Lemma 4.2, we know $\{v_k, v_j\}$ are coduplicates, and hence, rows (columns) j and k have the same entries, except possibly in two positions, namely $A_{k,k} = d_k$, $A_{j,j} = d_j$. Moreover, $A_{k,j} = A_{j,k} = 1$. By representing the rows j and k of the matrix M, we observe that the congruence operations

$$R_k \leftarrow R_k - R_j, \qquad C_k \leftarrow C_k - C_j$$

give

$$
\begin{array}{c} j \\ \\ k \end{array}
\begin{pmatrix} a_1 \ldots a_i \ldots d_j & 1 \ldots a_n \\ \cdot \quad \cdot \quad \cdot \quad \cdot \quad \cdot \\ a_1 \ldots a_i \ldots 1 & d_k \ldots a_n \end{pmatrix}
\longrightarrow
\begin{array}{c} j \\ \\ k \end{array}
\begin{pmatrix} a_1 \ldots a_i \ldots & d_j & 1-d_j & \ldots a_n \\ \cdot \quad \cdot \quad \cdot & \cdot & \cdot & \cdot \\ 0 \ldots 0 \ldots & 1-d_j & d_k + d_j - 2 \ldots 0 \end{pmatrix}.
$$

We observe that, together, the operations above preserve the symmetry of the matrix. This is why we omit the columns when depicting the resulting matrix.

After this, most of the nonzero elements in row and column k have been removed. But we must now remove the two entries $1 - d_j$. There are three subcases, depending on whether $d_k + d_l - 2 \neq 0$ and whether $d_j = 1$.

Subcase 1a: $d_k + d_j - 2 \neq 0$

Then we may perform the operations

$$R_j \leftarrow R_j - \frac{1-d_j}{d_k + d_j - 2} R_k, \qquad C_j \leftarrow C_j - \frac{1-d_j}{d_k + d_j - 2} C_k$$

obtaining

$$
\begin{array}{c} j \\ \\ k \end{array}
\begin{pmatrix} a_1 \ldots a_i \ldots \gamma \ldots & 0 & \ldots a_n \\ \cdot \quad \cdot \quad \cdot \quad \cdot & \cdot & \cdot \\ 0 \ldots 0 \ldots 0 \ldots & d_k + d_j - 2 \ldots 0 \end{pmatrix},
$$

where

$$\gamma = d_j - \frac{(1-d_j)^2}{d_k + d_j - 2} = \frac{d_k d_j - 1}{d_k + d_j - 2}.$$

The following assignments are made by the algorithm, reflecting the net result of these operations.

$$d_k \leftarrow d_k + d_j - 2, \qquad d_j \leftarrow \frac{d_k d_j - 1}{d_k + d_j - 2}. \tag{5.1}$$

As row (and column) k is diagonalized, the value d_k becomes permanent, and we may remove v_k from the cotree. We notice that this removal requires a rearrangement of the cotree, so that it remains minimal. Lemma 5.1 in the next section explains how.

$$T_G \leftarrow T_G - v_k. \tag{5.2}$$

Note that the assignments in (5.1) are technically incorrect since d_k is modified in the first assignment and used in the second. In our algorithm's pseudocode, we will use temporary variables. However, in order to keep notation simple, we omit temporary variables in the remainder of this section.

Subcase 1b: $d_k + d_j = 2$ **and** $d_j = 1$
In this case, rows k and j of the matrix look like

$$\begin{matrix} j \\ \cdot \\ k \end{matrix} \begin{pmatrix} a_1 \ldots a_i \ldots 1 \ldots 0 \ldots a_n \\ \cdot \quad \cdot \quad \cdot \quad \cdot \quad \cdot \\ 0 \ldots 0 \ldots 0 \ldots 0 \ldots 0 \end{pmatrix},$$

and we are done. We make the assignments

$$d_k \leftarrow 0, \qquad d_j \leftarrow 1, \qquad T_G \leftarrow T_G - v_k,$$

as d_k becomes permanent, and v_k is removed.

Subcase 1c: $d_k + d_j - 2 = 0$ **and** $d_j \neq 1$
Then our matrix looks like

$$\begin{matrix} j \\ \cdot \\ k \end{matrix} \begin{pmatrix} a_1 \ldots a_i \ldots \quad d_j \quad \ldots 1 - d_j \ldots a_n \\ \cdot \quad \cdot \quad \cdot \quad \cdot \quad \cdot \\ 0 \ldots 0 \ldots 1 - d_j \ldots \quad 0 \quad \ldots 0 \end{pmatrix}.$$

We notice that $a_1, \ldots, a_n \in \{0, 1\}$, and since $1 - d_j \neq 0$, for each $i \neq k$ such that $a_i = 1$, we perform

$$R_i \leftarrow R_i - \frac{1}{1-d_j} R_k, \qquad C_i \leftarrow C_i - \frac{1}{1-d_j} C_k. \tag{5.3}$$

This annihilates most entries in row (and column) j, without changing any other values, yielding

$$j \begin{pmatrix} 0 \ldots 0 \ldots & d_j & \ldots 1 - d_j \ldots 0 \\ \cdot & \cdot & \cdot & \cdot & \cdot \\ k & 0 \ldots 0 \ldots 1 - d_j \ldots & 0 & \ldots 0 \end{pmatrix}.$$

The operations

$$R_j \leftarrow R_j + \tfrac{1}{2} R_k, \quad C_j \leftarrow C_j + \tfrac{1}{2} C_k$$

replace the diagonal d_j with the value 1, while the operations

$$R_k \leftarrow R_k - (1 - d_j) R_j; \quad C_k \leftarrow C_k - (1 - d_j) C_j$$

eliminate the two off-diagonal elements, finally giving

$$j \begin{pmatrix} 0 \ldots 0 \ldots 1 \ldots & 0 & \ldots 0 \\ \cdot & \cdot & \cdot & \cdot & \cdot \\ k & 0 \ldots 0 \ldots 0 \ldots & -(1 - d_j)^2 & \ldots 0 \end{pmatrix}.$$

What is different about this subcase is that both rows k and j have been diagonalized. We make the following assignments:

$$d_k \leftarrow -(1 - d_j)^2, \qquad d_j \leftarrow 1, \qquad T_G \leftarrow T_G - v_k, \qquad T_G \leftarrow T_G - v_j,$$

removing both vertices from T_G and making both variables permanent.

Case 2: $w = \cup$
As v_j and v_k are duplicates, by Lemma 4.2, rows (and columns) k and j of M have the same values except possibly in two positions, namely, $A_{k,k} = d_k$, $A_{j,j} = d_j$. Here $A_{j,k} = A_{k,j} = 0$.
    Similar to Case 1, the row and column operations

$$R_k \leftarrow R_k - R_j, \quad C_k \leftarrow C_k - C_j$$

yield the following transformation:

$$j \begin{pmatrix} a_1 \ldots a_i \ldots d_j & 0 & \ldots a_n \\ \cdot & \cdot & \cdot & \cdot & \cdot \\ k & a_1 \ldots a_i \ldots 0 & d_k & \ldots a_n \end{pmatrix} \longrightarrow j \begin{pmatrix} a_1 \ldots a_i \ldots d_j & -d_j & \ldots a_n \\ \cdot & \cdot & \cdot & \cdot & \cdot \\ k & 0 \ldots 0 \ldots -d_j & d_k + d_j & \ldots 0 \end{pmatrix}.$$

Subcase 2a: $d_k + d_j \neq 0$
Then the matrix operations

$$R_j \leftarrow R_j + \frac{d_j}{d_k + d_j} R_k. \quad C_j \leftarrow C_j + \frac{d_j}{d_k + d_j} C_k$$

diagonalize the matrix, and the following assignments are made.

$$d_k \leftarrow d_k + d_j \qquad d_j \leftarrow \frac{d_k d_j}{d_k + d_j}, \qquad T_G \leftarrow T_G - v_k.$$

Subcase 2b: $d_k + d_j = 0$ **and** $d_j = 0$
As in the case **subcase 1b**, the row and column k of the matrix are in diagonal form, and we assign

$$d_k \leftarrow 0, \qquad d_j \leftarrow 0, \qquad T_G \leftarrow T_G - v_k.$$

Subcase 2c: $d_k + d_j = 0$ **and** $d_j \neq 0$
Since $d_j \neq 0$, we use row and column operations similar to (5.3) to annihilate most of entries in row j and column j:

$$\begin{array}{c} j \\ \\ k \end{array} \begin{pmatrix} 0 \ldots 0 \ldots & d_j & \ldots -d_j \ldots 0 \\ \cdot \quad\quad \cdot & \cdot & \cdot \quad\quad \cdot \\ 0 \ldots 0 \ldots -d_j & \ldots & 0 \ldots 0 \end{pmatrix}.$$

The congruence operations

$$R_k \leftarrow R_k + R_j, \quad C_k \leftarrow C_k + C_j$$

complete the diagonalization. The following assignments are made.

$$d_k \leftarrow -d_j, \qquad d_j \leftarrow d_j, \qquad T_G \leftarrow T_G - v_k, \qquad T_G \leftarrow T_G - v_j.$$

5.2   Diagonalizing $A + xI$

Based on the diagonalization of a row and a column described in the previous section, we provide an algorithm that constructs a diagonal matrix D, congruent to $M = A + xI$, where A is the adjacency matrix of a cograph G, and $x \in \mathbb{R}$. It takes as input the minimal cotree T_G, initializing all entries of D with x.

At the beginning of each iteration, the cotree represents the subgraph induced by vertices whose rows and columns have not yet been diagonalized. During each iteration, a pair of siblings $\{v_k, v_j\}$ from the cotree is selected, whose existence is guaranteed by Corollary 4.2. Note that, after the initial step, the algorithm cannot assume that either of the diagonal elements d_k and d_j are still x.

Each iteration of the loop annihilates either one or two rows (columns), updating diagonal values as described in Sect. 5.1. Once a row (column) is diagonalized, its entries never participate again in row and column operations, and so their values remain unchanged. When a row (column) corresponding to vertex v has been diagonalized, the subgraph induced by those vertices whose rows (columns) still need to be diagonalized has been reduced.

We then remove v from the cotree. Since the class of cographs is hereditary (see Sect. 4.1 and Theorem 4.2), if G is a cograph and $v \in V(G)$, the induced subgraph $G - v$ is also a cograph. There is a general method for constructing a (minimal) cotree associated with $G - v$ from a (minimal) cotree T_G associated with G, see [6, Lem. 1]. We consider this construction in the special case in which the cotree is in normalized form and the deleted vertex v is a leaf of maximal depth of T_G. Note that, in this case, vertex v must have a sibling that is a leaf of the same depth.

Lemma 5.1 *Let T_G be the minimal cotree of a cograph $G = (V, E)$ with $|V| \geq 2$, let $v, u \in V$ be siblings of greatest depth in T_G, and let w be their parent. Assume that w has $k \geq 2$ children. The following hold for the minimal cotree T_{G-v} of $G - v$:*

(a) *If $k > 2$, T_{G-v} is obtained by deleting v from T_G.*
(b) *If $k = 2$ and w is not the root of T_G, T_{G-v} is obtained by deleting v and w from T_G and by adding an edge between u and the parent of w in T_G.*
(c) *If $k = 2$ and w is the root, T_{G-v} is a singleton with root w.*

We omit the proof since this follows easily from the properties of the cotrees given in Sect. 4.2. We apply this procedure in every iteration, and we next illustrate this in a full example.

Figure 5.1 is the pseudocode of the algorithm `Diagonalize Cograph`. We notice that the cotree can be represented with a data structure whose size is $O(n)$, and since the algorithm does not store the matrix and does not actually perform the congruence operations described in Sect. 5.1, but rather only records changes on the diagonal values d_i of M, only $O(n)$ space is necessary. The algorithm terminates when all rows and columns have been diagonalized. Each iteration of `Diagonalize Cograph` takes constant time, so its running time is $O(n)$.

Example 5.1 We illustrate the application of the algorithm. As was the case for algorithm `Diagonalize Tree` (see Chap. 3), for small cographs, the algorithm may be performed by hand.

Consider the cograph G whose cotree is given in Fig. 5.2a. In this example, we consider the value $x = 1$, which means the algorithm outputs a diagonal matrix

```
Input: minimal cotree T_G, scalar x
Output: diagonal matrix D = diag(d_1, d_2, ..., d_n) congruent to A(G) + xI
Algorithm Diagonalize Cograph(T_G, x)
    initialize d_i := x, for 1 ≤ i ≤ n
    while T_G has ≥ 2 leaves
        select siblings {v_k, v_j} of maximum depth with parent w
        α ← d_k    β ← d_j
        if w = ⊕
            if α + β ≠ 2                    //subcase 1a
                d_k ← α + β − 2,  d_j ← (αβ−1)/(α+β−2),  T_G = T_G − v_k
            else if β = 1                    //subcase 1b
                d_k ← 0,  d_j ← 1,  T_G = T_G − v_k
            else                             //subcase 1c
                d_k ← −(1 − β)²,  d_j ← 1,  T_G = T_G − v_k,  T_G = T_G − v_j
        else if w = ∪
            if α + β ≠ 0                    //subcase 2a
                d_k ← α + β,  d_j ← αβ/(α+β),  T_G = T_G − v_k
            else if β = 0                    //subcase 2b
                d_k ← 0,  d_j ← 0,  T_G = T_G − v_k
            else                             //subcase 2c
                d_k ← −β,  d_j ← β,  T_G = T_G − v_k,  T_G = T_G − v_j
    end loop
```

Fig. 5.1 Diagonalizing $A(G) + xI$

that is congruent to $A(G) + I$. We initialize all diagonal values with $x = 1$. We choose to start with the red vertices of Fig. 5.2a. We have $w = \oplus$, $\alpha = \beta = 1$, and hence $\alpha + \beta = 2$, and this is subcase 1(b), meaning that we set $d_j = 1$ and that we produce a permanent value $d_k = 0$. Figure 5.2b represents the permanent (isolated vertex) with value $d_k = 0$ and the new value for $d_j = 1$ that is now a single leaf of node $\oplus$. This is no longer a minimal cotree so, by using Lemma 5.1, we reconstruct the minimal cotree, represented in Fig. 5.2c. The second iteration is made by choosing the red leaves of maximum depth. This iteration is like the first one, and thus, according to subcase 1(b), we output another permanent value of $d_k = 0$, and Fig. 5.2d shows the reconstructed cotree and the two red leaves chosen for the third iteration. Here, we have $w = \cup$ and $\alpha = \beta = 1$, and because $\alpha + \beta = 2 \neq 0$, we are in subcase 2(a), meaning the permanent value produced is $d_k = \alpha + \beta = 2$, whereas $d_j = \alpha\beta/(\alpha+\beta) = 1/2$. Figure 5.2e shows the reconstructed cotree and new siblings in red to be processed. We have $\alpha = 1$, $\beta = 1/2$. Thus subcase 2(a) produces $d_k = \alpha + \beta = 3/2$ and $d_j = \alpha\beta/(\alpha+\beta) = 1/3$. Figure 5.2f shows the reconstructed cotree and new siblings in red for the fifth iteration. Here $w = \oplus$ and $\alpha = 1$, $\beta = \frac{1}{3}$, and because $\alpha + \beta = \frac{4}{3} \neq 2$, we are in subcase 1(a), and the permanent value $d_k = \alpha + \beta - 2 = -2/3$ is outputted, while $d_j = \alpha\beta-1/(\alpha+\beta-2) = 1$. The final iteration between the siblings that remain is given in Fig. 5.2g. This iteration is like the first one, meaning that the two final values are

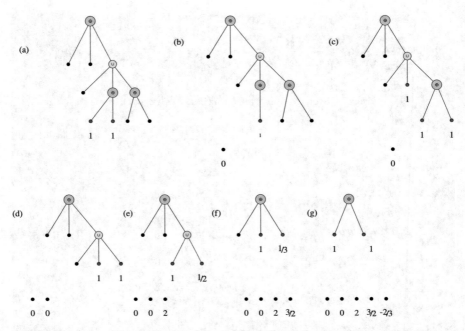

Fig. 5.2 Seven steps in an application of the algorithm Diagonalize Cographs; parts (a)–(g)

$d_k = 0$ and $d_j = 1$. The diagonal values of D are $0, 0, 2, 3/2, -2/3, 0, 1$. There is one negative value, and there are 3 positive values and 3 zero values. This implies that $x = -1$ is an eigenvalue of G with multiplicity 3, that 3 eigenvalues are greater than -1, and that just one is smaller than -1.

Example 5.2 If the same cotree is used with $x = 0$, we obtain three positive diagonal numbers and four negative diagonal numbers meaning that the inertia is $(3,4,0)$.

We observe that the algorithm presented here locates eigenvalues of adjacency matrices of cographs, but in fact the algorithm may be easily extended to locate eigenvalues of a slightly more general class of matrices. If a matrix M is any symmetric matrix whose underlying graph is a cograph, whose nonzero off-diagonal entries are all equal to some value w (equivalently, for the adjacency matrix of any weighted cograph whose edge weights are all equal to w), it is not difficult to see that the algorithm applies. It is necessary to initialize every value d_i as $M[i, i] + x$ in the algorithm. Additionally, in the elementary row and column operations of Sect. 5.1, we need to replace the matrix value 1 (which comes from the adjacency matrix) with the weight w.

5.3 Applications: Inertia and Spectral Characterization of Cographs

The algorithm presented in Sect. 5.2 is from 2018 and is a generalization of an algorithm given in 2013 [9] for locating eigenvalues in *threshold graphs*. This is an important hereditary subclass of cographs, namely the class of $\{P_4, C_4, 2K_2\}$-free graphs. It is shown in [14, Cor. 3.2] that the cotree of a threshold graph is a *caterpillar*, namely a tree that becomes a path when its leaves are removed.

It is interesting that in [10], it is shown that threshold graphs do not have eigenvalues in the interval $(-1, 0)$, which is somehow unexpected because the set of eigenvalues of graphs in general is *dense* in the real numbers: any nonempty open interval has an eigenvalue of some graph.

> In fact even more is true: any nonempty open interval of the real line contains an eigenvalue of a tree.

This is a result due to J. Salez [13]. The fact that $(-1, 0)$ has no eigenvalue of cographs was proved in [12], and this may be used to characterize cographs in terms of their spectra, as observed by E. Ghorbani [7]. The remainder of this section is devoted to prove this characterization, which will be done by using the algorithm `Diagonalize Cograph`. In the process, we obtain a formula for the inertia of cographs.

Theorem 5.1 *A graph G is a cograph if and only if no induced subgraph of G has an eigenvalue in* $(-1, 0)$.

We start with some technical lemmas.

Lemma 5.2 *Let G be a cograph with minimal cotree T_G. Let $\{v_k, v_j\}$ be a sibling pair processed by* `Diagonalize Cograph` *with parent w, for which* $0 \le d_k, d_j < 1$.

(a) *If $w = \oplus$, then d_k becomes permanently negative, and d_j is assigned a value in* $(0, 1)$.

(b) *If $w = \cup$, then d_k becomes permanently non-negative and d_j is assigned a value in* $[0, 1)$.

Proof Item (a). By our assumptions, **subcase 1a** is executed. Hence,

$$d_k \leftarrow \alpha + \beta - 2,$$

$$d_j \leftarrow \frac{\alpha\beta - 1}{\alpha + \beta - 2},$$

where α, β are the old values of d_k, d_j. Clearly, $d_k < 0$. Now $d_j > 0$, as both numerator and denominator are negative. Since $(\alpha\beta - 1) - (\alpha + \beta - 2) = (\alpha - 1)(\beta - 1) > 0$, it follows that $d_j - 1 = \frac{(\alpha\beta-1)-(\alpha+\beta-2)}{\alpha+\beta-2} < 0$ or that $d_j < 1$.

For item (b), we may have $d_k = d_j = 0$, and we execute **subcase 2b**; hence, d_k and d_j are assigned 0. In any other case, we execute **subcase 2a** and then

$$d_k \leftarrow \alpha + \beta,$$

$$d_j \leftarrow \frac{\alpha\beta}{\alpha + \beta},$$

where α, β are the old values of d_k, d_j. Clearly, $d_k > 0$. As for d_j, we observe that $d_j \geq 0$, since the denominator is positive and the numerator is non-negative. Since $(\alpha\beta) - (\alpha + \beta) = (\alpha - 1)(\beta - 1) - 1 < 0$, it follows that $d_j < 1$. $\square$

We state a technical lemma with a slight change in the range of the initial values. The proof may be done in a similar way as the proof of the previous lemma.

Lemma 5.3 *Let G be a cograph with minimal cotree T_G. Let $\{v_k, v_j\}$ be a sibling pair processed by* Diagonalize Cograph *with parent w, for which $0 < d_k, d_j \leq 1$.*

(a) *If $w = \oplus$, then d_k becomes permanently nonpositive and $d_j \in (0, 1]$.*
(b) *If $w = \cup$, then d_k becomes permanently positive and $d_j \in (0, 1)$.*

Lemma 5.4 *Let G be a cograph with minimal cotree T_G. Consider the execution of* Diagonalize Cograph(T_G, x).

(a) *If $x = 0$, then all diagonal values of vertices remaining on the cotree are in $[0, 1)$.*
(b) *If $x = 1$, then all diagonal values of vertices remaining on the cotree are in $(0, 1]$.*

Proof As both proofs are similar, we prove item (a) and omit the proof of item (b). Initially, all values on T_G are zero. Suppose after m iterations of Diagonalize Cograph all diagonal values of the cotree are in $[0, 1)$, and consider iteration $m + 1$ with sibling pair $\{v_k, v_j\}$ and parent w. By assumption, $0 \leq d_k, d_l < 1$. If $w = \oplus$, then Lemma 5.2 (a) guarantees the vertex d_j remaining on the tree is assigned a value in $(0, 1)$. If $w = \cup$, Lemma 5.2 (b) guarantees $d_j \in [0, 1)$. This means after $m + 1$ iterations the desired property holds, completing the proof by induction. $\square$

We recall that when a cograph G is disconnected, then its minimal cotree T_G has root of type $\cup$ and, in particular, it has $t \geq 0$ leaves representing the isolated vertices of G.

Remark 5.1 Observe that if w is an internal node in T_G having t children, as the algorithm progresses bottom-up through the rules of Lemma 5.1, each internal child of w eventually is replaced by a leaf. Thus when w is ready to be processed, it will have t leaves as children. To simplify our analysis, without loss of generality, we can assume that all $t - 1$ sibling pairs are processed consecutively.

Remark 5.2 Consider a node w of type $\cup$ with $\ell = s + t$ children, where s children are internal (of type $\oplus$) and t children are leaves. In the execution of `Diagonalize Cograph`($T_G, 0$), from Lemmas 5.2 (a) and 5.4 (a), each $\oplus$ node will become positive. Thus when w is processed, it will have s leaves with positive values and t leaves with zero.

Proposition 5.1 *Let G be a cograph with minimal cotree T_G having $t \geq 0$ isolated vertices. Assume T_G has k nodes of type $\oplus$ denoted by $\{w_1, w_2, \ldots, w_k\}$ and ℓ nodes of type $\cup$ denoted by $\{w_{k+1}, w_{k+2}, \ldots, w_{k+\ell}\}$. Let $q_i = s_i + t_i$ be the number of children of w_i, where s_i is the number of internal children and t_i is the number of leaves of w_i, $i = 1, \ldots, k + \ell$. Let U be the set of $\cup$ nodes that have two or more leaves, that is, those in which $t_i \geq 2$. Then the number $n_-(G)$ of negative eigenvalues of G, and the multiplicity $m_{A(G)}(0)$ of 0 as an eigenvalue of G, is given by*

$$ n_-(G) = \sum_{i=1}^{k}(q_i - 1); \tag{5.4} $$

$$ m_{A(G)}(0) = \sum_{w_i \in U} (t_i - 1) + \delta(t), \quad \text{where } \delta(t) = \begin{cases} 1 & \text{if } t > 0 \\ 0 & \text{if } t = 0 \end{cases}. \tag{5.5} $$

Proof To prove Eq. 5.4, we execute `Diagonalize Cograph`($T_G, 0$) and consider any node w_i in T_G of type $\oplus$ with $q_i = s_i + t_i$ children, where t_i are leaves and s_i are internal nodes of type $\cup$. By Remark 5.1, when w_i is eligible to be processed, it will have q_i leaves as children and will process $q_i - 1$ sibling pairs. By Lemma 5.4 (a), all diagonal values on the cotree remain in $[0, 1)$. By Lemma 5.2 (a), each of the $q_i - 1$ sibling pairs will produce a permanent negative value before w_i is removed. This shows $n_-(G) \geq \sum_{i=1}^{k}(q_i - 1)$. However, Lemma 5.2 (b) shows that processing a sibling pair with parent $\cup$ can only produce non-negative permanent values. Hence, the inequality is tight, completing the proof for $n_-(G)$.

To prove Eq. 5.5, let w_i be an internal node of type $\cup$ having $q_i = s_i + t_i$ children where s_i children are internal nodes and t_i are leaves. By Remark 5.2, when w_i is ready to be processed, it will have $q_i = s_i + t_i$ leaves, s_i positive values, and t_i zeros. For every pair of zero values, we execute **subcase 2b** of algorithm `Diagonaliza Cograph`. Each execution of **subcase 2b** produces a permanent zero value. This shows that w_i contributes with $t_i - 1$ zeros to the diagonal matrix, whenever $t_i > 1$. Additionally, we see from the proof of Lemma 5.2 (b) that the remaining permanent values produced while processing w_i are positive. This shows $m_G(0) \geq \sum_{w_i \in U}(t_i - 1)$. To obtain the value for $m_G(0)$, we first note that, by Lemma 5.2 (a), no zero can be created when processing a sibling pair whose parent is $\oplus$. If the cograph is connected, then we are done as $t = 0$ and the root is $\oplus$.

If G is disconnected, then the root of T_G has type $\cup$ with $s+t$ children, with $t \geq 0$ leaves and $s \geq 0$ internal nodes of type $\oplus$. We can assume $s > 0$, for otherwise the graph is a collection of isolated vertices. We claim that exactly t additional zeros are created. Indeed, when the root is processed, all the t leaves have value zero and the s internal nodes became leaves with positive values by Lemma 5.2 (a). From Lemma 5.2 (b), we apply $s - 1$ times **subcase 2a**, creating $s - 1$ permanent positive values and leaving a positive value in the cotree. Then $t - 1$ zero permanent values are created through **subcase 2b**. The last iteration, when **subcase 2a** is applied, creates an additional zero. $\square$

Corollary 5.1 *The inertia of the n vertex cograph G, having the same parameters of Proposition 5.1, is given by the triple*

$$(p, q, r),$$

where $q = \sum_{i=1}^{k}(q_i - 1), \quad r = \sum_{w_i \in U}(t_i - 1) + \delta(t)$ *and* $p = n - q - r$.

Proposition 5.2 *Let G be a cograph with minimal cotree T_G. Assume T_G has k nodes of type $\oplus$ denoted by $\{w_1, w_2, \ldots, w_k\}$ and ℓ nodes of type $\cup$ denoted by $\{w_{k+1}, w_{k+2}, \ldots, w_{k+\ell}\}$. Let $q_i = s_i + t_i$ be the number of children of w_i, where s_i is the number of internal nodes and t_i is the number of leaves of w_i, $i = 1, \ldots, k+\ell$. Let J be the set of $\oplus$ nodes having two or more leaves, that is, those for which $t_i \geq 2$. Then the number $m_{A(G)}(-1, \infty)$ of eigenvalues greater than -1 and the multiplicity of -1 as an eigenvalue of G are given by*

$$m_{A(G)}(-1, \infty) = 1 + \sum_{i=k+1}^{k+\ell} (q_i - 1), \tag{5.6}$$

$$m_{A(G)}(-1) = \sum_{w_i \in J} (t_i - 1). \tag{5.7}$$

Proof To show Eq. (5.7), we count the number of zeroes in the diagonal matrix outputted by the execution of `Diagonalize Cograph`$(T_G, 1)$. Let w_i be a node of type $\oplus$ with $q_i = s_i + t_i$ children, $t_i \geq 0$ leaves, and $s_i \geq 0$ internal nodes of type $\cup$. As in the previous result, when w_i is processed, all the s_i internal nodes became leaves with positive values smaller than 1, by Lemma 5.3 (b). Whenever $t_i \geq 2$, we notice that $t_i - 1$ permanent zero values are produced, as we are in the case **subcase 1b** of the algorithm. Moreover, any other pair processed by node w_i produces no zero value by Lemma 5.2 (a). This means that $m_{A(G)}(-1) \geq \sum_{w_i \in J}(t_i - 1)$. To see that equality holds, we first notice that, by Lemma 5.4(b), all values that remain to be processed are in $(0, 1]$. Since these values will be processed by $\cup$ nodes, we see by Lemma 5.3 (b) that no additional zeroes are produced. This proves Eq. (5.7).

To show Eq. (5.6), we count the number of positive values in the diagonal matrix outputted by the execution of `Diagonalize Cograph`$(T_G, 1)$. Consider now a node w_i of type $\cup$ with $q_i = s_i + t_i$ children, where t_i are leaves and s_i are nodes of type $\oplus$, which eventually become leaves with positive values (by Lemma 5.4 (b)). Then all values of the q_i leaves are in $(0,1]$; by Lemma 5.3 (b), $q_i - 1$ positive values are outputted; hence, this shows that $m_{A(G)}(-1, \infty) \geq \sum_{i=k+1}^{k+\ell} (q_i - 1)$.

We claim that the final iteration of the algorithm always outputs an additional positive value. To see this, let $q = s + t \geq 2$ be the number of leaves of the root. We can assume that $s > 0$ for otherwise G is either the complete graph (if the root is $\oplus$) or a collection of isolates (if the root is $\cup$). Let d_j and d_k be the last two values in the cotree. By Lemma 5.4 (b), $d_j, d_k \in (0, 1]$. If the root is $\cup$, then the final iteration executes **subcase 2a**, which outputs two positive values. One of them is already accounted for in the formula $\sum_{i=k+1}^{k+\ell} (q_i - 1)$.

If the root is $\oplus$, since $s > 0$, we may assume $d_j \in (0, 1)$ by Lemma 5.3 (b). Now the value d_k is in $(0, 1)$, when $t = 0$, while $d_k = 1$, when $t > 0$. In any case, the last iteration executes **subcase 1a**, which produces a positive value and a negative value. Therefore, we have $m_{A(G)}(-1, \infty) \geq 1 + \sum_{i=k+1}^{k+\ell} (q_i - 1)$.

To show equality, we observe that, by Lemma 5.3 (a), no positive value is outputted by processing two siblings of a node $\oplus$ with values in $(0,1]$.   □

We remark that the formulas for $m_{A(G)}(0)$ and $m_{A(G)}(-1)$ appear in Bıyıkoğlu, Simić and Stanić [4, Cor. 3.2]. We presented here an alternate proof using our algorithm.

Proof of Theorem 5.1 Let G be a cograph. The number $m_{A(G)}(-1, 0)$ of eigenvalues in $(-1, 0)$ is given by

$$m_{A(G)}(-1, 0) = m_{A(G)}(-1, \infty) - m_{A(G)}[0, \infty)$$
$$= m_{A(G)}(-1, \infty) - n_+(G) - m_{A(G)}(0)$$
$$= m_{A(G)}(-1, \infty) - (n - n_-(G) - m_{A(G)}(0)) - m_{A(G)}(0)$$
$$= m_{A(G)}(-1, \infty) - n + n_-(G).$$

Now, by Eq. (5.4) and Eq. (5.6), it follows that

$$m_{A(G)}(-1, 0) = 1 + \sum_{i=k+1}^{k+\ell} (q_i - 1) - n + \sum_{i=1}^{k} (q_i - 1)$$
$$= 1 + \sum_{i=1}^{k+\ell} (q_i - 1) - n = 0.$$

By Lemma 2.3, the number of leaves in the minimal cotree is $n = 1 + \sum_{i=1}^{k+\ell} (q_i - 1)$ and the last equality follows. This shows that no cograph has eigenvalue in $(-1, 0)$. As cographs are an hereditary class (see Chap. 4) of graphs, any induced subgraph of G is also a cograph, and hence, any induced subgraph of G has no eigenvalue in $(-1, 0)$.

Conversely, if G is not a cograph, then G has an induced P_4 whose eigenvalues are approximately ± 1.61803, ∓ 0.61803; hence, P_4 has an eigenvalue in $(-1, 0)$.

$\square$

We end this chapter describing other possible applications of the algorithm to research problems that are relevant in matrix and graph theory.

It is quite interesting that cographs do not have eigenvalues in $(-1, 0)$ and, in fact, that this essentially characterizes this class of graphs. Are there other natural graph classes whose set of eigenvalues is not dense in the real line? Which open intervals do not contain eigenvalues of graphs in such a class? In a recent development, the authors of [3] proved that, for a given scalar N, there are infinitely many threshold graphs having no eigenvalues in $(0, N]$. The method uses the algorithm described in this chapter, which was also used to find infinitely many threshold graphs having no eigenvalues in $[-N, -1)$. It takes advantage of the fact that the cotrees associated with threshold graphs are caterpillars.

Other problems that may be studied by the eigenvalue location method presented in this chapter lie in the core of spectral graph theory. They involve finding *integral graphs*—graphs whose eigenvalues are composed only by integers [2], finding cospectral mates, and computing eigenvalue multiplicities [1]. Additionally, the method has been used to study graph energy, a parameter given by the sum of absolute values of the eigenvalues [1, 11].

We now turn to problems related to the spectral radius of matrices. In a 1972 paper, Hoffman [8] posed the problem of finding limit points of the spectral radii of symmetric matrices with non-negative entries. In the same paper, Hoffman showed that it is sufficient to consider symmetric matrices having entries in $\{0, 1\}$ and zero diagonal, that is, adjacency matrices of graphs. The set L of Hoffman limit points contains the real numbers α for which there exists a sequence $\{G_k\}$ of graphs such that $\rho(G_i) \neq \rho(G_j)$, for all $i \neq j$, and $\lim_{k \to \infty} \rho(G_k) = \alpha$, where $\rho(G)$ denotes the spectral radius of a graph G. In his original paper, Hoffman fully determined the intersection $L \cap [0, (2 + \sqrt{5})^{1/2}]$. Since then, mathematicians are in pursuit of what is called the *Hoffman Program*, which consists of identifying elements $\alpha \in L$ and of characterizing the connected graphs whose spectral radius does not exceed such a limit point α [16]. This program includes related questions. For instance, given a specific class of symmetric matrices $\mathcal{M}$, say Laplacian matrices of graphs, find the subset of Hoffman limit points $L_{\mathcal{M}}$ defined using only the spectral radius of matrices in this class. Similarly, given a specific class of graphs C, consider the subset of Hoffman limit points L_C defined for matrices associated with graphs in C. This raises natural questions for cographs. For instance, do there exist real numbers that are limit points of spectral radii of cographs? The second component of the Hoffman program is related to the Hoffman–Smith limit points discussed at the end of Chap. 3. Characterizing classes of cographs whose spectral radii do not exceed a certain fixed number seems a feasible goal using eigenvalue location methods.

We conclude with a well-known problem that was first posed by Brualdi and Hoffman in 1976 and asks for the connected graph with n vertices and m edges having the largest spectral radius. This problem has a long history (see, for example, [15]), but it has not been settled to this day. Brualdi and Solheid [5] have shown that the nonzero entries of the adjacency matrix of such a graph must have a shape that comes from threshold graphs. It is very tempting to believe that our method could be used to determine which threshold graph is the right candidate.

References

1. Allem, L.E., Tura, F.C.: Multiplicity of eigenvalues of cographs. Discrete Appl. Math. **247**, 43–52 (2018). https://doi.org/10.1016/j.dam.2018.02.010
2. Allem, L.E., Tura, F.: Integral cographs. Discrete Appl. Math. **283**, 153–167 (2020). https://doi.org/10.1016/j.dam.2019.12.021
3. Allem, L.E., Oliveira, E.R., Tura, F.: On i-eigenvalue free threshold graphs (2021). https://doi.org/10.48550/ARXIV.2110.12107. https://arxiv.org/abs/2110.12107
4. Bıyıkoğlu, T., Simić, S.K., Stanić, Z.: Some notes on spectra of cographs. Ars Combin. **100**, 421–434 (2011)
5. Brualdi, R., Solheid, E.S.: On the spectral radius of connected graphs. Publ. Inst. Math. Belgrade **39**(53), 45–54 (1986)
6. Corneil, D.G., Lerchs, H., Burlingham, L.S.: Complement reducible graphs. Discrete Appl. Math. **3**(3), 163–174 (1981). https://doi.org/10.1016/0166-218X(81)90013-5
7. Ghorbani, E.: Spectral properties of cographs and p5-free graphs. Linear Multilinear Algebra **67**(8), 1701–1710 (2019). https://doi.org/10.1080/03081087.2018.1466865
8. Hoffman, A.J.: On limit points of spectral radii of non-negative symmetric integral matrices. In: Y. Alavi, D.R. Lick, A.T. White (eds.) Graph Theory and Applications, pp. 165–172. Springer, Berlin, Heidelberg (1972)
9. Jacobs, D.P., Trevisan, V., Tura, F.: Eigenvalue location in threshold graphs. Linear Algebra Appl. **439**(10), 2762–2773 (2013). https://doi.org/10.1016/j.laa.2013.07.030
10. Jacobs, D.P., Trevisan, V., Tura, F.: Eigenvalues and energy in threshold graphs. Linear Algebra Appl. **465**, 412–425 (2015). https://doi.org/10.1016/j.laa.2014.09.043
11. Jacobs, D.P., Trevisan, V., Tura, F.: Eigenvalue location in cographs. Discrete Appl. Math. **245**, 220–235 (2018). https://doi.org/10.1016/j.dam.2017.02.007
12. Mohammadian, A., Trevisan, V.: Some spectral properties of cographs. Discrete Math. **339**(4), 1261–1264 (2016). https://doi.org/10.1016/j.disc.2015.11.005
13. Salez, J.: Every totally real algebraic integer is a tree eigenvalue. J. Combin. Theory B **111**, 249–256 (2015). https://doi.org/10.1016/j.jctb.2014.09.001
14. Sciriha, I., Farrugia, S.: On the spectrum of threshold graphs. ISRN Discrete Math. **2011**, 1–29 (2011). https://doi.org/10.5402/2011/108509
15. Stevanović, D.: Spectral Radius of Graphs. Academic Press (2014)
16. Wang, J., Wang, J., Brunetti, M.: The Hoffman program of graphs: old and new (2020). https://doi.org/10.48550/ARXIV.2012.13079. https://arxiv.org/abs/2012.13079

Chapter 6
Locating Eigenvalues Using Tree Decomposition

6.1 Gaussian Elimination and Tree Decompositions

From their origin, the tree decompositions of Sect. 4.3 have been connected to *Gaussian elimination*, particularly in the situation where it is performed on a sparse matrix. As is well known, Gaussian elimination is a basic step in classical algorithms for many problems in linear algebra, such as solving linear equations and computing the rank, the determinant, and the inverse of a matrix. It is also closely related to the notion of equivalent matrices described in Sect. 2.2, in that the input matrix is equivalent to the matrix obtained by this procedure.

Let $A = (a_{ij})$ denote an $n \times m$ matrix. We say that A is in *row echelon form* (also called *row canonical form*) if the following hold:

(a) If $j < i$, then $a_{ij} = 0$.
(b) The *pivot* of row i is the element $a_{ij} \neq 0$ with least j, if it exists. If rows $i_1 < i_2$ have pivots in columns j_1 and j_2, respectively, then $j_1 < j_2$.

The following are matrices in row echelon form:

$$\begin{pmatrix} 1 & 0 & 1 & -1 & 2 \\ 0 & 0 & 2 & -1 & 0 \\ 0 & 0 & 0 & 0 & -1 \end{pmatrix}, \quad \begin{pmatrix} 1 & -1 & 1 \\ 0 & 1 & 2 \\ 0 & 0 & 0 \end{pmatrix}.$$

With the terminology of Sect. 2.4, Gaussian elimination is a procedure to transform any matrix M into a matrix in row echelon form using elementary row operations. In our description, we focus on elementary row operations of types I and III. Each step in the process consists of a sequence of elementary operations of type III where multiples of some row R_i, whose first nonzero element M_{ij} lies in column j, are added to other rows so that the other nonzero elements in column j become 0 and M_{ij} becomes a pivot. Elementary row operations of type I are used to ensure that the pivots in different rows respect the order prescribed in part (b) of the definition

© The Author(s), under exclusive license to Springer Nature Switzerland AG 2022
C. Hoppen et al., *Locating Eigenvalues in Graphs*, SpringerBriefs in Mathematics,
https://doi.org/10.1007/978-3-031-11698-8_6

above. To speed up this procedure, one would like to transform M into a matrix in row echelon form with a small number of field operations.[1] A *pivoting scheme* (or *elimination ordering*) determines the order in which the rows are used to produce new pivots. Formally, it may be viewed as an order on the rows of the matrix with the property that the first row according to this order is used in the first step to produce the leftmost pivot, the second row is used in the second step to produce the second pivot from the left, and so on, until all rows either contain pivots or are equal to the zero vector. In a first course in linear algebra, the pivoting scheme is typically defined during the elimination process itself. Starting with a matrix M_0, the first row is chosen among the rows that contain the leftmost nonzero entry, which becomes the first pivot. We perform elementary row operations to eliminate the other nonzero entries in its column, which produces a matrix M_1. The second row is chosen among the rows with the leftmost nonzero entry in the modified matrix, except for the first row, and so on. This ensures that later steps do not change earlier work.

When M is a sparse matrix, we may turn it into row echelon form with a small number of operations if we find a pivoting scheme with small *fill-in*, defined as the set of matrix positions that were initially 0, but became nonzero at some point during the computation. A possible problem with using a predefined pivoting scheme is that accidental cancelations may occur (an *accidental cancelation* happens when a nonzero element becomes zero owing to an operation aimed at canceling a different element in the same row). We should mention that, despite seeming to be helpful at first sight, accidental cancelations may mess up the pivoting scheme, as the unexpected zero could be the element that was originally prescribed as a pivot, and cannot be a pivot anymore. In the previous paragraph, we did not worry about accidental cancelations because the pivoting scheme was not set up in advance. However, we neither took the fill-in into account, and the sparsity of the input matrix could be lost. Pivoting schemes based on the structure of the matrix have been an important tool to perform Gaussian elimination preserving sparsity and have been exploited in several algorithms, including treewidth-based algorithms [1, 5, 6]. To get some intuition on why the treewidth may be helpful in this context, we come back to our discussion at the beginning of Sect. 4.3, where we mentioned that if an n-vertex graph G has a tree decomposition of width k, then all of the adjacencies of G may be identified with $O(nk)$ queries. Recall that a query consists of asking, for a given pair $\{u, v\} \in V(G)$, whether $\{u, v\}$ is an edge of G. Since the edges of the underlying graph of a matrix M correspond to the off-diagonal nonzero entries of the matrix, assuming that there are no accidental cancelations, this implies that the number of nonzero elements that need to be eliminated by the algorithm to turn the matrix into row echelon form is $O(kn)$. Further assume that we may define a pivoting scheme with the following properties: (i) whenever the pivoting scheme designates row R_i for the next pivot, all the elements that need to be eliminated are in rows whose corresponding vertices lie in a bag B that also contains i, the

---

[1] As in previous chapters, we shall focus on the field of real numbers, but this discussion would be valid for matrices with entries in any field.

vertex that corresponds to it in the graph; (ii) whenever a multiple of a row R_i is added to a row R_j, then all nonzero elements in R_i lie in columns corresponding to vertices in B. This means that multiples of row R_i will be added at most k times to other rows. Moreover, each such addition will alter at most k entries of the row R_j being modified by this operation, leading to $O(k^2)$ operations to create a pivot in row R_i, and hence, to $O(k^2 n)$ operations overall. This unfortunately does not work in general due to accidental cancelations, but the papers cited above have used variations of this idea to devise efficient algorithms for Gaussian elimination.

In this chapter, our aim is not to describe an algorithm for Gaussian elimination, but to find an algorithm that, given any symmetric matrix M, computes a diagonal matrix D that is congruent to it. The structure of the matrix is captured by a nice tree decomposition of the graph G associated with it (see Sect. 4.4), which in turn induces a pivoting scheme that makes the algorithm particularly fast if G has small width. Of course, accidental cancelations may happen, and we will need to deal with them. Observe that this has already been done in the algorithms of Chaps. 3 and 5. To be specific, this happens when a child v_j of v_k has $d_j = 0$ in the algorithm of Fig. 3.1. In the algorithm of Fig. 5.1, accidental cancelations occur in some form for all subcases except 1a and 2a.

The following theorem formally states the main conclusion of this chapter.

Theorem 6.1 *Given an $n \times n$ symmetric matrix M and a tree decomposition $\mathcal{T}$ of width k for its underlying graph, algorithm* Diagonalize Treewidth *(see Sect. 6.2) produces a diagonal matrix D congruent to M in time $O(k \log k |\mathcal{T}| + k^2 n)$.*

It is important to notice that, as it is usually the case for algorithms that make use of a tree decomposition, we assume that the tree decomposition $\mathcal{T}$ is given as part of the input. We also observe that the term $O(k \log k |\mathcal{T}|)$ corresponds to the cost of computing a nice tree decomposition of width k starting from $\mathcal{T}$ (see Lemma 4.3), so that the algorithm runs in time $O(k^2 n)$ assuming that the tree decomposition is already a nice tree decomposition of width k. Moreover, we assume that we have a compact representation of the input matrix M because, in standard form, just reading M would already take quadratic time. A possible representation could be a list of triples (i, j, m_{ij}) containing all nonzero entries. For convenience, we assume that each node i in the nice tree decomposition that forgets a vertex v records all values m_{uv} such that $u \in B_i \cup \{v\}$, which means that we need to store a vector of length at most $k + 1$ in each forget node, whose entries are in the original matrix. This representation can be obtained efficiently from the list representation, while the nice tree decomposition is constructed.

We also note that, but for a constant factor, the bound obtained in Theorem 6.1 is best possible for any diagonalization algorithm based on elementary row and column operations. Finally, we would like to point out that determining whether Gaussian elimination may be performed using $O(k^2 n)$ operations is a significant open problem. While writing this book, the best known algorithm, devised by Fomin et al. [1], runs in time $O(k^3 n)$.

6.2 Diagonalization Algorithm

In this section, we describe the algorithm `Diagonalize Treewidth`. This algorithm has been first presented in [3], and an earlier version may be found at [2]. Let M be an $n \times n$ symmetric matrix, and let $G = (V, E)$ be the underlying graph with vertex set $V = [n]$ associated with it. Each vertex of G corresponds to a row (and column) of M, so that we shall often refer to a row or column of M as a row or column of a vertex $v \in V(G)$. We wish to find a diagonal matrix congruent to M. Note that this algorithm may be used to locate the eigenvalues of any symmetric matrix M with underlying graph G, such as the adjacency and the Laplacian matrices of G, for instance. Given a real number x, we may simply use $M - xI$ as an input and look at the number of positive, negative, and zero entries in the diagonal of the diagonal matrix obtained by the algorithm.

As part of the input, we assume that we are given a nice tree decomposition $\mathcal{T}$ of G of width k. Actually, Lemma 4.3 tells us how to turn any tree decomposition of G into a nice tree decomposition, so that it would be possible to start with an arbitrary tree decomposition of G of width k. The algorithm works bottom-up in the rooted tree $\mathcal{T}$, so we order the nodes $1, \ldots, m$ of $\mathcal{T}$ in postorder and operate on a node i after its children have been processed. We note that in our description, we do not refer to any particular data structure, and instead assume that we have oracle access to the nice tree decomposition and to the original entries of M, as described in Sect. 4.3.

High-Level Description of the Algorithm We start with a high-level description of the algorithm, which is summarized in Fig. 6.1. Each node i in the tree decomposition, except the root, produces a pair of matrices $(N_i^{(1)}, N_i^{(2)})$, which may be combined into a symmetric matrix N_i with at most $2(k + 1)$ rows. The rows and columns of $N_i^{(1)}$ and $N_i^{(2)}$ are associated with rows and columns of the input matrix M. We shall refer to N_i as the *box* produced by node i. When referring to a box, we shall assume that the rows and columns are labeled by vertices of M.

```
Input: a nice tree decomposition T of the underlying
graph G associated with M of width k and the nonzero entries of M
Output: diagonal matrix D = diag(d₁,...,dₙ) congruent to M

Diagonalize Treewidth(M)
order the nodes of T as 1, 2,...,m in postorder
for i from 1 to m do
    if i is a Leaf then (Nᵢ⁽¹⁾, Nᵢ⁽²⁾)=LeafBox(Bᵢ)
    if i introduces vertex v and has child j then (Nᵢ⁽¹⁾, Nᵢ⁽²⁾)=IntroduceBox(Bᵢ, v, Nⱼ)
    if i is a Join with children j and ℓ then (Nᵢ⁽¹⁾, Nᵢ⁽²⁾)=JoinBox(Bᵢ, Nⱼ, Nₗ)
    if i forgets vertex v and has child j then (Nᵢ⁽¹⁾, Nᵢ⁽²⁾)=ForgetBox(Bᵢ, v, Nⱼ)
end loop
```

Fig. 6.1 High-level description of the algorithm `Diagonalize Treewidth`

As in the other algorithms in previous chapters, the algorithm traverses the tree from the leaves to the root so that, at node i, the algorithm either initializes the pair $(N_i^{(1)}, N_i^{(2)})$ or it produces $(N_i^{(1)}, N_i^{(2)})$ based on the matrices transmitted by its children and local information extracted from M. It then transmits the pair to its parent. In each step, the algorithm may also produce diagonal elements of a matrix congruent to M. These diagonal elements are not transmitted by a node to its parent but are appended to a global array. At the end of the algorithm, the array consists of the diagonal elements of a diagonal matrix D that is congruent to M.

Step i of the algorithm refers to the iteration of the loop of the algorithm corresponding to node i; we say that it *processes* node i. The box N_i produced by a node on the tree has the form

$$
N_i = \begin{array}{|c|c|}
\hline
N_i^{(0)} & N_i^{(1)} \\
\hline
N_i^{(1)T} & N_i^{(2)} \\
\hline
\end{array} , \tag{6.1}
$$

where $N_i^{(0)} = \mathbf{0}_{k_i' \times k_i'}$, $N_i^{(2)}$ is a $k_i'' \times k_i''$ symmetric matrix, and $N_i^{(1)}$ is a $k_i' \times k_i''$ matrix in row echelon form such that every row has a pivot. In particular, this ensures that $k_i' \leq k_i''$. Moreover, the width of the tree decomposition will ensure that $k_i'' \leq k+1$, leading to $0 \leq k_i' \leq k_i'' \leq k+1$. Observe that k_i' can be zero, in which case we regard $N_i^{(0)}$ and $N_i^{(1)}$ as empty. An important feature of these matrices is that each row of N_i (and the corresponding column) is associated with a vertex of G, and therefore with a row (and column) of M. Let $V(N_i)$ denote the set of vertices of G associated with the rows of N_i. We say that the k_i' rows in $N_i^{(0)}$ have *type I* and the k_i'' rows of $N_i^{(2)}$ have *type II*. This is represented by the partition $V(N_i) = V_1(N_i) \cup V_2(N_i)$, where $V_1(N_i)$ and $V_2(N_i)$ have type I and type II, respectively. Note that $N_i^{(1)}$ has rows of type I and columns of type II. An important feature of the algorithm is that the vertices of type II are precisely the vertices in the bag B_i from the tree decomposition associated with node i. This is summarized by the following lemma.

Lemma 6.1 *For all $i \in [m]$, the box N_i defined in terms of the pair $(N_i^{(1)}, N_i^{(2)})$ produced by node i satisfies the following properties:*

(a) $0 \leq k_i' \leq k_i'' \leq k + 1$.

(b) $N_i^{(1)}$ is a matrix in row echelon form such that every row has a pivot.

(c) $N_i^{(2)}$ is symmetric and satisfies $V_2(N_i) = B_i$, with rows and columns sorted in increasing order.

To have an intuition about how the algorithm works, consider that we are trying to apply the strategy for Gaussian elimination described in Sect. 6.1 on M, but performing the *same* column operation after each row operation to obtain a matrix that is congruent to the original matrix. The pivoting scheme that we choose is the following. Recall that each row of M is associated with a vertex v of G, which in turn is associated with the unique Forget node of the nice tree decomposition $\mathcal{T}$ that forgets v. Given a bottom-up ordering of $\mathcal{T}$, the first row in the pivoting scheme is the row associated with the first Forget node in this ordering, the second row is the one associated with the second Forget node, and so on. The algorithm attempts to diagonalize row v at the node where v is forgotten. If the diagonal entry corresponding to v in the box is nonzero, we use it to annihilate nonzero elements in this row/column, which turns out to lead to a new pivot. At this point, its row and column will be fully diagonalized and will not to be considered ever again by the algorithm. If we think in terms of the box, vertices in $V(N_i)$ correspond to rows that have yet to be diagonalized at the end of step i of the algorithm.[2] Vertices of type II are rows that will come later in the pivoting scheme, while vertices v of type I are rows that have already appeared in the pivoting scheme, but which have not been diagonalized because of an accidental cancelation. Rows of type I may be viewed as a temporary buffer, whose size is kept under control because k' satisfies $k' \leq k'' \leq k + 1$. In the process of producing boxes N_i where $N_i^{(1)}$ is in row echelon form, some of these rows become diagonalized.

Regarding the description in Fig. 6.1, any row v of the input matrix M begins as a type II row after one or more steps of type LeafBox or IntroduceBox. It can either be diagonalized during the application of ForgetBox to the node of $\mathcal{T}$ that forgets v, or it becomes a type I row at this step, and finally becomes diagonalized in a later step of type JoinBox or ForgetBox. JoinBox and ForgetBox are the only steps in which the algorithm actually performs row and column operations, which can produce new diagonal elements. ForgetBox is the only step in which a row of type II may become type I. In JoinBox, some type I rows may be diagonalized, but rows cannot change types.

Relating the Algorithm with Operations in the Original Matrix In this book, we will not give a rigorous proof of the correctness of algorithm `Diagonalize Treewidth`, which may be found in [3]. Our aim is to guide the reader through the motivation behind each step of the algorithm. Before describing each step in detail, we discuss how the content of the boxes relates with the original matrix and the diagonalization process. To record the diagonal entries produced by the algorithm, let D_i be the array containing all pairs (v, d_v), where v is a vertex of G and d_v is the diagonal entry associated with it, produced up to the end of step i. Let $\tilde{M}(i)$ denote the matrix obtained from M by performing all row and column operations that the algorithm has performed in any box up to the end of step i, so that by

---

[2] It is possible that some of these rows and columns have already been diagonalized, but that the algorithm has yet to identify this.

construction $\tilde{M}(i)$ is congruent to M.[3] Even though the algorithm does not compute $\tilde{M}(i)$, the reason why it works is that, at the end of the algorithm, the array D_m contains exactly one pair for each $v \in V$ and $\tilde{M}(m) = diag(d_1, \ldots, d_n)$, so that $diag(d_1, \ldots, d_n)$ is congruent to M.

It turns out that, for all $i \in [m]$, the rows that are actually computed in N_i coincide with the entries of $\tilde{M}(i)$ for type I rows. More precisely, for a type I row u in N_i and any column v in the matrix, the entries uv and vu in $\tilde{M}(i)$ and N_i coincide, if $v \in V(N_i)$, and the entries uv and vu in $\tilde{M}(i)$ are equal to 0, if $v \notin V(N_i)$. This is consistent with the intuition that rows of type I have already been partially diagonalized and lie in a temporary buffer, in the sense that all the remaining nonzero entries lie in a small matrix N_i. However, this connection does not hold in general for type II rows. Indeed, the entries of $N_i^{(2)}$ can only capture changes due to operations performed while processing the descendants of i in $\mathcal{T}$, but vertices of type II can be simultaneously in many different branches that merge together higher up in the tree. To deal with this, the algorithm will not "see" all the nonzero entries of M from the start, and the nonzero entry corresponding to a pair $\{u, v\}$ will only be taken into account at the node that forgets u or v (whichever happens first). In short, any entry uv of N_i where u and v have type II only keeps track of the variation in the entry uv of M because of elementary operations performed in steps j where j is a node in the subtree of $\mathcal{T}$ rooted i.

To motivate some of the steps of the algorithm, it is convenient to state some more facts about it, for which we need additional notation. At the start of the algorithm, let M_0 be the input matrix M. As the algorithm proceeds bottom-up on $\mathcal{T}$, we need to understand how it acts on its subtrees. Let $\mathcal{T}_i$ be the subtree of $\mathcal{T}$ rooted at node i. Given $v \in V(G)$, let $\mathcal{T}(v)$ denote the subtree of $\mathcal{T}$ induced by all nodes whose bag contains v. We consider $\mathcal{T}(v)$ to be rooted at the child j of the node i that forgets v. Note that, among all nodes whose bag contains v, j is the closest to the root of $\mathcal{T}$. To see why this is true, recall that the root has an empty bag. So, if ℓ is any node such that $v \in B_\ell$, then the (single) node that forgets v must be along the path that connects ℓ to the root.

Let M_i be the symmetric matrix obtained from M as follows, where for clarity, we use the notation $Q[u, v]$ to denote the entry Q_{uv} of a matrix Q:

$$M_i[u, v] = \begin{cases} m_{uv}, & \text{if } \mathcal{T}(u) \cap \mathcal{T}_i \neq \emptyset, \mathcal{T}(v) \cap \mathcal{T}_i \neq \emptyset, \{u, v\} \nsubseteq B_i, \\ 0, & \text{otherwise.} \end{cases} \quad (6.2)$$

By definition, this means that M_i coincides with M at all pairs (u, v) such that each of u and v appears in the bag of some node of $\mathcal{T}_i$, and such that at most one of u and v lies in B_i. In other words, at least one of them has already been forgotten. Intuitively, this means the entries of M_i are the entries that have already been seen

---

[3] Since the rows and columns of the boxes are indexed by vertices of G, the same operations can easily be performed on the entire matrix. They are performed in the same order as in the algorithm.

by the algorithm while acting on nodes of $\mathcal{T}_i$. Since the root of $\mathcal{T}_m = \mathcal{T}$ has an empty bag and all vertices appear in some bag of the tree decomposition $\mathcal{T}$, we eventually have $M_i = M$. This always holds for $i = m$.

Let $\tilde{M}_i$ be the matrix that is congruent to M_i obtained by performing all row and column operations performed by the algorithm after processing the nodes in the branch $\mathcal{T}_i$. We emphasize that we only keep track of the matrices M_i and $\tilde{M}_i$ (and of the matrix $\tilde{M}(i)$ defined above) to justify why the algorithm works, and they are not actually computed by the algorithm.

Our main invariants control the relationship between N_i and the diagonal elements produced in step i with the matrices M_i, $\tilde{M}_i$ and $\tilde{M}(i)$. The lemma below states that some intuitive facts about the algorithm hold. Part (a) mentions the simple fact that matrices obtained from one another by a sequence of pairs of the same elementary row and column operations are congruent. Part (b) establishes that any pairs added to the global array D_i stay there and that each row is associated with at most one diagonal element. Part (c) states the important fact that a row can only be added to another row in step i if it becomes diagonalized or has type I at the end of the step. For convenience, we let $\pi_1(D_i)$ and $\pi_2(D_i)$ denote the projections of D_i onto their first and second coordinates, respectively, so that $\pi_1(D_i)$ is the set of rows that have been diagonalized up to the end of step i and $\pi_2(D_i)$ is the (multi)set of diagonal elements found up to this step.

Lemma 6.2 *The following facts hold for all $i \in [m]$.*

(a) $\tilde{M}(i)$ and $\tilde{M}_i$ are symmetric matrices congruent to M and M_i, respectively.
(b) $D_{i-1} \subseteq D_i$ and $\pi_1(D_i)$ is injective.
(c) If a multiple of row (or column) v has been added to a row (or column) u in step i, then $v \in \pi_1(D_i \setminus D_{i-1}) \cup V_1(N_i)$.

The next lemma relates subtrees of type $\mathcal{T}(v)$ and $\mathcal{T}_i$ with vertex types. Part (a) states that any row v that has been diagonalized during step i or is associated with a row of box N_i must be contained in a bag of a node in the branch $\mathcal{T}_i$. For the rows of part (a), part (b) further specifies that, unless v has type II, the subtree $\mathcal{T}(v)$ is fully contained in $\mathcal{T}_i$. It also states that, for any row w that was diagonalized in earlier steps, either $\mathcal{T}(w)$ is a subtree of $\mathcal{T}_i$ or is completely disjoint from $\mathcal{T}_i$. Conversely, part (c) tells us that any vertex that lies in a bag of a node of $\mathcal{T}_i$ is in $V(N_i)$ or was diagonalized in earlier steps.

Lemma 6.3 *The following facts hold for all $i \in [m]$.*

(a) If $v \in \pi_1(D_i \setminus D_{i-1}) \cup V(N_i)$, then $\mathcal{T}_i \cap \mathcal{T}(v) \neq \emptyset$.
(b) If $v \in \pi_1(D_i) \cup V_1(N_i)$, then $\mathcal{T}_i \cap \mathcal{T}(v) = \emptyset$ or $\mathcal{T}(v)$ is a subtree of $\mathcal{T}_i$.
(c) Let v be such that $\mathcal{T}(v) \cap \mathcal{T}_i \neq \emptyset$. Then $v \in V(N_i) \cup \pi_1(D_i)$.

Lemmas 6.2 and 6.3 may be proved by induction on i, starting from the trivial case $i = 0$. After each step, we need to show that the properties hold for i assuming that they hold for smaller values.

```
Input: a set B_i of size b_i
Output: a box N_i = (N_i^(1), N_i^(2))

LeafBox(B_i)
    Set N_i^(1) = ∅
    N_i^(2) = 0_{b_i×b_i} .
```

Fig. 6.2 Procedure LeafBox

```
Input: a node i with bag B_i, child j, B_j = B_i − v, and a box N_j = (N_j^(1), N_j^(2))
Output: a box N_i = (N_i^(1), N_i^(2))

IntroduceBox(B_i, v, N_j)
    N_i^(1) = N_j^(1),  N_i^(2) = N_j^(2)
    add zero row and zero column associated with v to N_i^(2)
        making sure that type-ii rows remain in increasing order
    if N_i^(1) is nonempty, add zero column v to N_i^(1)
        preserving the column order of N_i^(2)
```

Fig. 6.3 Procedure IntroduceBox

Detailed Description of the Algorithm We now describe each step of Algorithm Diagonalize Treewidth. When the node is a leaf corresponding to a bag B_i of size b_i, we apply procedure LeafBox (Fig. 6.2). This procedure initializes a box N_i to be transmitted up the tree in a very simple way. It is such that $k' = 0$ and $k'' = b_i$, where $N_i^{(2)} = \mathbf{0}_{k'' \times k''}$. By construction, the box N_i defined by LeafBox satisfies the properties of Lemma 6.1. Since no operations have been performed, if the statements of Lemmas 6.2 and 6.3 hold up to the end of step $i - 1$, and step i processes a leaf i, then they must also hold at the end of step i.

Next, we explain the procedure associated with a node of type Introduce. The inputs are the set B_i, the vertex v that has been introduced, and the box $N_j = (N_j^{(1)}, N_j^{(2)})$ obtained after processing the child B_j of B_i. The box N_i is obtained from N_j by adding a new type II row/column corresponding to vertex v (this becomes the last row/column of the matrix), whose elements are all zero. It is obvious that the matrix N_i produced by IntroduceBox (Fig. 6.3) satisfies the properties of Lemma 6.1. In both cases, no row/column operation is performed, $\tilde{M}(i) = \tilde{M}(i - 1)$, $\tilde{M}_i = \tilde{M}_j$, and $D_i = D_{i-1}$. As before, if the statements of Lemmas 6.2 and 6.3 hold up to the end of step $i-1$, and step i processes an introduce node i, then they must also hold at the end of step i.

Input: a node i with bag B_i and boxes N_j, N_ℓ associated with its two children
Output: a box $N_i = (N_i^{(1)}, N_i^{(2)})$

JoinBox(B_i, N_j, N_ℓ)
   construct N_i^* as in (6.3)
   do row and column operations as in (6.4) on $N_i^{(1)*}$ to achieve row echelon form
   for each zero row of $N_i^{(1)*}$ (indexed by a vertex u), add $(u, 0)$ to D_i
   $N_i^{(1)}$ is $N_i^{(1)*}$ with zero row/columns removed as in (6.5)

Fig. 6.4 Procedure JoinBox

We now address the operation associated with nodes of type Join. Let i be a node
of type Join, and let N_j and N_ℓ be the matrices transmitted by its children, where
$j < \ell < i$. By Lemma 6.1(c) and the definition of the Join operation, we have
$V_2(N_j) = V_2(N_\ell)$. We claim that $V_1(N_j) \cap V_1(N_\ell) = \emptyset$. To see why this is true,
let $u \in V_1(N_j)$ and $v \in V_1(N_\ell)$. By Lemma 6.3(a) and (b) (which may be applied
inductively), $\mathcal{T}(u)$ and $\mathcal{T}(v)$ are subtrees of $\mathcal{T}_j$ and $\mathcal{T}_\ell$, respectively, so $u \neq v$.

The JoinBox operation (Fig. 6.4) first creates a matrix N_i^* whose rows and
columns are labeled by $V_1(N_j) \cup V_1(N_\ell) \cup V_2(N_j)$ with the structure below. Assume
that $|V_1(N_j)| = r$, $|V_1(N_\ell)| = s$, and $|B_i| = t$, and define

$$N_i^* = \begin{array}{|c|c|c|} \hline \mathbf{0}_{r \times r} & \mathbf{0}_{r \times s} & N_j^{(1)} \\ \hline \mathbf{0}_{s \times r} & \mathbf{0}_{s \times s} & N_\ell^{(1)} \\ \hline N_j^{(1)T} & N_\ell^{(1)T} & N_i^{*(2)} \\ \hline \end{array}, \qquad (6.3)$$

where $N_i^{*(2)} = N_j^{(2)} + N_\ell^{(2)}$. Let $N_i^{*(0)} = \mathbf{0}_{(r+s) \times (r+s)}$ denote the matrix on the top
left corner. Note that the matrix

$$N_i^{*(1)} = \begin{array}{|c|} \hline N_j^{(1)} \\ \hline N_\ell^{(1)} \\ \hline \end{array}$$

is an $(r + s) \times t$ matrix consisting of two matrices in row echelon form on top of
each other. The intuition behind this definition of N_i^* is as follows. Since $V_1(N_j) \cap$
$V_1(N_\ell) = \emptyset$, the entries uv such that $u \in V_1(N_j)$ and $v \in V_1(N_\ell)$, or vice versa, are
equal to 0 in the partially diagonalized matrix (see Lemma 6.4(b) below). Regarding
$N_i^{*(2)}$, first note that the matrix M_i defined in (6.2) satisfies $M_i = M_j + M_\ell$. Also,
under the assumption that the entries of $N_j^{(2)}$ and $N_\ell^{(2)}$ tell us the variation of the
corresponding entry of M due to elementary operations performed in nodes of $\mathcal{T}_j$
and $\mathcal{T}_\ell$, respectively, their sum gives the combined variation due to the elementary
operations in both branches.

To obtain N_i from N_i^*, we perform row operations on $N_i^{*(1)}$ followed by column operations with the same labels $N_i^{*(1)T}$. The goal is to insert the rows labeled by $V_1(N_\ell)$ (the *right rows*) into the matrix $N_j^{(1)}$ labeled by $V_1(N_j)$ (the *left rows*) to produce a single matrix in row echelon form with no zero rows. We do this by attempting to insert the last row w of $N_\ell^{(1)}$ into $N_j^{(1)}$. If the pivot of row w lies in a column c and no row of $N_j^{(1)}$ has a pivot in this column, we simply augment $N_j^{(1)}$ by inserting the row of w in the proper position (this is done by exchanging rows) and remove this row from $N_\ell^{(1)}$. As in the other steps, this is followed by the insertion of the last column of $N_\ell^{(1)T}$ as a column of $N_j^{(1)T}$ in its proper position. However, if the pivot α_c of row w lies in a column c and row v of $N_j^{(1)}$ also has a pivot β_c in column c, then we use v to eliminate the pivot of w by performing the operations

$$R_w \leftarrow R_w - \frac{\alpha_c}{\beta_c} R_v, \ C_w \leftarrow C_w - \frac{\alpha_c}{\beta_c} C_v \qquad (6.4)$$

in N_i^*. Note that the entries in $N_i^{*(0)}$ and $N_i^{*(2)}$ are not affected by these operations. Now either the row associated with w is a zero vector or it has a pivot in a row $c' > c$, and we try to insert it once again. We iterate this until all right rows have either been inserted into $N_j^{(1)}$ or have become zero vectors (and the same happened to the columns of $N_\ell^{(1)T}$). Note that if $r + s > k''$, we will certainly produce at least one zero row in $N_i^{*(1)}$. If Z_i denotes the set of vertices associated with rows whose entries are all zero, where $|Z_i| = z$, we let $D_i = D_{i-1} \cup \{(v, 0) : v \in Z_i\}$, and we remove the rows and columns associated with vertices in Z_i from $N_i^{*(1)}$ and $N_i^{*(1)T}$, respectively, to produce the matrix

$$N_i = \begin{array}{|c|c|} \hline \mathbf{0}_{k' \times k'} & N_i^{(1)} \\ \hline N_i^{(1)T} & N_i^{(2)} \\ \hline \end{array}, \qquad (6.5)$$

where $k' = r + s - z$, $k'' = t$, and $N_i^{(1)}$ is a $k' \times k''$ matrix in row echelon form and $N_i^{(2)} = N_i^{*(2)}$. We observe that N_i satisfies the properties of Lemma 6.1. Items (b) and (c) are satisfied by construction. For (a), the inequality $k' \leq k''$ is a consequence of the fact that $N_i^{(1)}$ is in row echelon form and has a pivot in every row, while $k'' \leq k+1$ follows from $k'' = |B_i|$, proving that Lemmas 6.2 and 6.3 hold after step i uses induction and the properties discussed above.

To conclude the description of the algorithm, we describe ForgetBox (Fig. 6.5). Assume that i forgets vertex v and let j be its child, so that $B_i = B_j \setminus \{v\}$. Since $\mathcal{T}(v)$ is rooted at j, it is a subtree of $\mathcal{T}_j$. This procedure starts with N_j and produces a new matrix N_i so that the row associated with v becomes of type I or is diagonalized. Since we now know that v will not appear in the bag of any node that will be processed later in the algorithm, we start by *introducing the entries* of M involving v in B_i. This means defining a new matrix N_i^* from the box N_j, where all entries of types uv and vu where $u \in B_j$ are replaced by $N_j^{(2)}[u, v] + m_{uv}$, while the other entries remain unchanged. This is the step where the algorithm "sees" the entry uv of M. We argue that this happens exactly once for each pair $\{u, v\}$ such that $m_{uv} \neq 0$. Part (2) of the definition of tree decomposition ensures that there is a bag B_j associated with a node j that contains both u and v. On the path from j to the root, there is a first node i such that $\{u, v\} \not\subseteq B_i$, so that node i forgets u or forgets v by the structure of a nice tree decomposition. This node i is the same for every j with this property because of part (3) in the definition of tree decomposition, and assume that i forgets u. At this point, $v \in B_i$, and the entry will be updated as above (and added to the matrix M_i in (6.2)). Note that vertex v will be forgotten at a node ℓ higher up the tree, but at that point $u \notin B_\ell$, so that the entries uv and vu will not be considered then.

We observe that in the box N_j the row corresponding to v is type II, so after introducing the entries of M, v is associated with a row in $N_i^{*(2)}$. For convenience,

```
Input: a node i with bag B_i, child j with B_i = B_j \ {v} and box N_j
Output: a box N_i = (N_i^(1), N_i^(2))

ForgetBox(B_i, v, N_j)
    N_i^*(1) = N_j^(1),  N_i^*(2)[u, w] = N_j^(2)[u, w], for all u, w ∈ B_j with v ∉ {u, w}
    N_i^*(2)[u, v] = N_i^*(2)[v, u] = N_j^(2)[u, v] + m_uv, for all u ∈ B_j
    exchange rows/columns so that N_i has the form of (6.6)
    if x_v is empty or 0 then
        if y_v is empty or 0 then      //subcase 1a
            add (v, d_v) to D
            remove row v from N_i
        else if d_v = 0                //subcase 1b
            do row/column operations as in (6.7)
            if row v gets a pivot then insert the row into N_i^(1)
            else add (v, 0) to D and remove row from N_i
        else // Here {d_v ≠ 0, y_v ≠ 0}.          //subcase 1c
            use d_v to diagonalize row/column v as in (6.8)
            add (v, d_v) do D and remove row v form N_i
    else // Here x_v ≠ 0.            //case 2
        let u be the vertex of the rightmost nonzero entry of x_v
        if x_v has other nonzero entries then
            eliminate them with the elementary operations (6.9)
        if d_v ≠ 0 then perform the elementary operations (6.10)
        perform the elementary operations (6.11)
        use d_v and d_u to diagonalize rows/columns v and u as in (6.7)
        add (v, d_v) and (u, d_u) to D and eliminate rows v and u from N_i.
```

Fig. 6.5 Procedure ForgetBox

after exchanging rows and columns, we look at N_i^* in the following way:[4]

$$
N_i^* = \begin{array}{c|c|c|c}
 & d_v & \mathbf{x}_v & \mathbf{y}_v \\
\hline
\mathbf{x}_v^T & \mathbf{0}_{k' \times k'} & N_i^{*(1)} \\
\hline
\mathbf{y}_v^T & N_i^{*(1)T} & N_i^{*(2)}
\end{array}.
$$

(6.6)

Here, the first row and column represent the row and column in N_i^* associated with v, while $N_i^{*(1)}$ and $N_i^{*(2)}$ determine the entries uw such that u has type I and $w \in B_i$, and such that $u, w \in B_i$, respectively. In particular, $\mathbf{x}_v$ and $\mathbf{y}_v$ are row vectors of sizes k_j' and $k_j'' - 1$, respectively.

Depending on the vectors $\mathbf{x}_v$ and $\mathbf{y}_v$, we proceed in different ways.

Case 1: $\mathbf{x}_v$ *is empty or* $\mathbf{x}_v = [0 \cdots 0]$.

If $\mathbf{y}_v = [0 \cdots 0]$ (or $\mathbf{y}_v$ is empty), the row of v is already diagonalized, and we simply add (v, d_v) to D_i and remove the row and column associated with v from N_i^* to produce N_i. We refer to this as subcase 1(a).

If $\mathbf{y}_v \neq [0 \cdots 0]$, there are again two options. In subcase 1(b), we assume that $d_v = 0$. The aim is to turn v into a row of type I. To do this, we need to insert $\mathbf{y}_v$ into the matrix $N_i^{*(1)}$ in a way that the resulting matrix is in row echelon form. Note that this may be done by only adding multiples of rows of $V(N_i^{*(1)})$ to the row associated with v. At each step, if the pivot α_j of the (current) row associated with v is in the same position of the pivot β_j of R_u, the row associated with vertex u already in $N_i^{*(1)}$, we use R_u to eliminate the pivot of R_v:

$$
R_v \leftarrow R_v - \frac{\alpha_j}{\beta_j} R_u, \quad C_v \leftarrow C_v - \frac{\alpha_j}{\beta_j} C_u.
$$

(6.7)

This is done until the pivot of the row associated with v may not be canceled by pivots of other rows, in which case the row associated with v may be inserted in the matrix (to produce the matrix $N_i^{(1)}$), or until the row associated with v becomes a zero row, in which case $(v, 0)$ is added to D_i, and we remove the row and column associated with v from N_i^* to produce N_i.

If $d_v \neq 0$, we are in subcase 1(c), and we use d_v to eliminate the nonzero entries in $\mathbf{y}_v$ and diagonalize the row corresponding to v. For each element $u \in B_i$ such that the entry α_v of $\mathbf{y}_v$ associated with u is nonzero, we perform

$$
R_u \leftarrow R_u - \frac{\alpha_v}{d_v} R_v, \quad C_u \leftarrow C_u - \frac{\alpha_v}{d_v} C_v.
$$

(6.8)

---

[4] This is helpful for visualizing the operations, but this step is not crucial in an implementation of this procedure.

When all such entries have been eliminated, we add (d_v, v) to D_i, and we let N_i be the remaining matrix. Observe that, in this case, $N_i^{(1)} = N_i^{*(1)}$, only the elements of $N_i^{*(2)}$ are modified to generate $N_i^{(2)}$.

Case 2: $\mathbf{x}_v$ is nonempty and $\mathbf{x}_v \neq [0 \cdots 0]$.

Let u be the vertex associated with the rightmost nonzero entry of x_v. Let α_j be this entry. We use this element to eliminate all the other nonzero entries in x_v, from right to left. Let w be the vertex associated with an entry $\alpha_\ell \neq 0$. We perform

$$R_w \leftarrow R_w - \frac{\alpha_\ell}{\alpha_j} R_u, \quad C_w \leftarrow C_w - \frac{\alpha_\ell}{\alpha_j} C_u. \tag{6.9}$$

A crucial fact is that the choice of u ensures that, even though these operations modify $N_i^{*(1)}$, the new matrix is still in row echelon form and has the same pivots as $N_i^{*(1)}$. If $d_v \neq 0$, we still use R_u to eliminate this element:

$$R_v \leftarrow R_v - \frac{d_v}{2\alpha_j} R_u, \quad C_v \leftarrow C_v - \frac{d_v}{2\alpha_j} C_u. \tag{6.10}$$

At this point, the only nonzero entries in the $(k' + 1) \times (k' + 1)$ left upper corner of the matrix obtained after performing these operations are in positions uv and vu (and are equal to α_j). We perform the operations

$$R_u \leftarrow R_u + \frac{1}{2} R_v, \ C_u \leftarrow C_u + \frac{1}{2} C_v, \ R_v \leftarrow R_v - R_u, \ C_v \leftarrow C_v - C_u. \tag{6.11}$$

The relevant entries of the matrix are modified as follows:

$$\begin{pmatrix} 0 & \alpha_j \\ \alpha_j & 0 \end{pmatrix} \rightarrow \begin{pmatrix} 0 & \alpha_j \\ \alpha_j & \alpha_j \end{pmatrix} \rightarrow \begin{pmatrix} -\alpha_j & 0 \\ 0 & \alpha_j \end{pmatrix}. \tag{6.12}$$

We are now in the position to use the diagonal elements to diagonalize the rows associated with v and u, as was done in Case 1, when $x_v = [0, \ldots, 0]$ and $d_v \neq 0$. At the end of the step, we add $(v, -\alpha_j)$ and (u, α_j) to D_i.

Proof of Theorem 6.1 The proof of Theorem 6.1 follows from Lemmas 6.2 and 6.3 and a third result, which relates the entries of the matrices N_i and the array D produced by the algorithm with the entries of $\tilde{M}(i)$ and $\tilde{M}_i$. It may also be proved by induction on i. Part (a) tells us that the entries of $N_i^{(2)}$ record the changes on M due to operations performed for nodes in the branch $\mathcal{T}_i$. Part (b) tells us that, for rows of type I in N_i, all nonzero entries in the partially diagonalized matrix lie in $N_i^{(1)}$. Part (c) states that the rows that we called "diagonalized" have indeed been diagonalized in the partially diagonalized matrix.   □

Lemma 6.4 *The following facts hold for all $i \in [m]$:*

(a) *If $u, v \in B_i$, then the entry uv of $\tilde{M}_i$ is equal to the entry uv of $N_i^{(2)}$.*

(b) *If $v \in V_1(N_i)$, then the row (and column) associated with v in $\tilde{M}_i$ coincides with the row (and column) associated with v in $\tilde{M}(i)$. For any $u \in V(G)$, the entries uv and vu are equal to 0 if $u \notin B_i$ and are equal to the corresponding entries in N_i if $u \in B_i$.*

(c) *If $(v, d_v) \in D_i$, then row (and column) associated with v in $\tilde{M}(i)$ consists of zeros, with the possible exception of the vth entry, which is equal to d_v. If $\mathcal{T}_i \cap \mathcal{T}(v) \neq \emptyset$, then the row (and column) associated with v in $\tilde{M}_i$ satisfies the same property.*

Before concluding this section, we discuss the running time of `Diagonalize Treewidth`. By Lemma 4.3, given an arbitrary tree decomposition of G of width k with m nodes, a nice tree decomposition $\mathcal{T}'$ of the same width k with fewer than $4n - 1$ nodes can be computed in time $O((k \log k)m + kn)$. Recall that the number of nodes of each type in $\mathcal{T}'$ is at most n.

We consider the number of operations performed at each type of tree node. `LeafBox` initializes a matrix in $O(k^2)$ trivial steps. `IntroduceBox` uses $O(k)$ steps. For `JoinBox` and `ForgetBox`, the main cost comes from row and column operations. As the matrices have at most $k + 1$ rows and columns, each such row or column operation requires $O(k)$ sums and multiplications. Regarding `ForgetBox`, when v is forgotten, either v is turned into a type I vertex, or its row and column, and possibly the row and column of another vertex u, are diagonalized. The latter requires at most $O(k)$ row and column operations. If v is turned into a type I vertex, then inserting it into the matrix $N_i^{(1)}$ in row echelon form takes at most $k + 1$ row operations. `JoinBox` can be most time-consuming. To insert just one row vector into a $k \times k$ matrix in row echelon form, and preserving this property by adding multiples of one vector to another, can require up to $k + 1$ row operations. Each such operation can be done in time $O(k)$. To combine two matrices in row echelon form into one such matrix, up to $k + 1$ row vectors are inserted. Thus the total time for this operation is $O(k^3)$. This immediately results in an upper bound of $O((k \log k)m + k^3 n)$ for the whole computation, where the first term can be omitted if we assume that we start with a compact nice tree decomposition, or with an arbitrary tree decomposition with $O(n)$ nodes for which the vertices in each bag are sorted according to a pre-determined order.

To realize that $O(k^3 n)$ may be replaced by $O(k^2 n)$, we have to employ a different accounting scheme for the time spent to keep $N_i^{(1)}$ in row echelon form in all applications of `JoinBox` and `ForgetBox`. We notice that every row operation in any $N_i^{(1)}$ creates at least one 0 entry in some row, meaning that its pivot moves at least one position to the right or creates a zero row. For every vertex v, its row is included at most once into some $N_i^{(1)}$, namely when v is forgotten. Its pivot can move at most k times, and the row becomes a row of 0 at most once. Thus considering all n vertices of G, at most $(k + 1)n$ row operations can happen for all matrices of type $N_i^{(1)}$. This only uses time $O(k^2 n)$. Thus, while during a single join procedure, $\Omega(k^2)$ row operations might be needed, the average is $O(k)$ such operations per join procedure.

6.3 Example

In this section, we show how the algorithm of the previous section acts on a concrete example. To this end, we consider the matrix

$$
M = (m_{ij}) = \begin{pmatrix} 1 & 1 & 1 & 0 & -1 \\ 1 & 0 & 0 & 2 & 0 \\ 1 & 0 & 1 & -1 & 0 \\ 0 & 2 & -1 & 1 & 0 \\ -1 & 0 & 0 & 0 & -1 \end{pmatrix},
$$

whose graph is depicted in Fig. 6.6, along with a nice tree decomposition. Note that this is the tree described in Fig. 4.5.

Suppose that we want to find the number of eigenvalues greater than 0 (and equal to and less than 0). We apply our algorithm with $c = 0$, that is, originally $M - cI = M$.

Assume that we have ordered the nodes of the tree in postorder so that the first three nodes are in the upper branch of Fig. 6.6, followed by the three nodes on the lower branch and by the four nodes starting from the node of type join. When we start, node 1 calls LeafBox with the bag of vertices $\{1, 3, 4\}$ producing the box $N_1 = \mathbf{0}_{3\times 3}$. This matrix has no type I vertices and three type II vertices, that is, $N_1^{(0)} = \emptyset$, $N_1^{(1)} = \emptyset$ and $N_1^{(2)} = N_1$.

The node $i = 2$, which forgets vertex 3, receives the box N_1 and produces $N_2^{*(2)}$ by adding entries m_{33}, m_{i3}, and m_{3j} of the original matrix M, where $i, j \in \{1, 4\}$, to the corresponding entries in $N_1^{(2)}$. Since the operations of ForgetBox deal with vertex $v = 3$, we exchange rows and columns so that 3 becomes the first

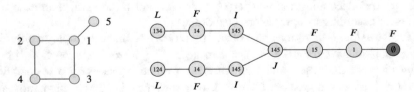

Fig. 6.6 The underlying graph of matrix M and a nice tree decomposition associated with it. The root of the tree is on the right

row/column, getting a matrix in the format of (6.6):

$$N_2^* = N_2^{*(2)} = \begin{matrix} 3 \\ 1 \\ 4 \end{matrix} \left(\begin{array}{c:cc} 1 & 1 & -1 \\ \hdashline 1 & 0 & 0 \\ -1 & 0 & 0 \end{array} \right).$$

The numbers on the left of the matrix denote the rows of M that are associated with each row (and the corresponding column) of $N_2^{*(2)}$. Using the terminology of ForgetBox, $\mathbf{x}_v$ is empty and $\mathbf{y}_v = [1, -1] \neq [0, 0]$. Since $d_v = 1$, we are in subcase 1c, and the algorithm diagonalizes the row/columns corresponding to v. To this end, we perform the operations $R_1 \leftarrow R_1 - R_3$,[5] followed by $C_1 \leftarrow C_1 - C_3$, producing the matrix

$$A_1 = \begin{matrix} 3 \\ 1 \\ 4 \end{matrix} \left(\begin{array}{ccc} 1 & 0 & -1 \\ 0 & -1 & 1 \\ -1 & 1 & 0 \end{array} \right),$$

followed by the operations $R_4 \leftarrow R_4 + R_3$ and $C_4 \leftarrow C_4 + C_3$, giving

$$A_2 = \begin{matrix} 3 \\ 1 \\ 4 \end{matrix} \left(\begin{array}{c:cc} 1 & 0 & 0 \\ \hdashline 0 & -1 & 1 \\ 0 & 1 & -1 \end{array} \right).$$

We have diagonalized the row corresponding to vertex 3. Hence, the node $i = 2$ on the tree adds $(v, d_v) = (3, 1)$ to the array D and transmits the box N_2 given by $N_2^{(0)} = N_2^{(1)} = \emptyset$ and

$$N_2^{(2)} = \begin{matrix} 1 \\ 4 \end{matrix} \left(\begin{array}{cc} -1 & 1 \\ 1 & -1 \end{array} \right)$$

to its parent. The node $i = 3$ introduces vertex 5, producing the box N_3 given by $N_3^{(0)} = N_3^{(1)} = \emptyset$ and

$$N_3 = N_3^{(2)} = \begin{matrix} 1 \\ 4 \\ 5 \end{matrix} \left(\begin{array}{ccc} -1 & 1 & 0 \\ 1 & -1 & 0 \\ 0 & 0 & 0 \end{array} \right).$$

Working similarly on the lower branch, the node $i = 4$ initiates a zero matrix $N_4 = \mathbf{0}_{3 \times 3}$ indexed by $\{1, 2, 4\}$ and transmits it to its parent, which forgets vertex

---

[5] When performing row and column operations, we shall always refer to the vertices that label each row and column.

$v = 2$. Following the instructions of ForgetBox, we define $N_5^{*(2)}$ as in (6.6):

$$N_5^{*(2)} = \begin{matrix} 2 \\ 1 \\ 4 \end{matrix} \begin{pmatrix} 0\ 1\ 2 \\ 1\ 0\ 0 \\ 2\ 0\ 0 \end{pmatrix}.$$

Thus we have $\mathbf{x}_v$ empty, $\mathbf{y}_v = [1, 2]$, and $d_v = 0$. We are in subcase 1b, and the algorithm asks us to insert the row indexed by 2 to $N_5^{(1)}$ and to perform operations, if necessary, to keep $N_5^{(1)}$ in row echelon form. Since this matrix was empty before adding the row, no operations are required. We arrive at

$$N_5 = \begin{matrix} 2 \\ 1 \\ 4 \end{matrix} \begin{pmatrix} 0\ 1\ 2 \\ 1\ 0\ 0 \\ 2\ 0\ 0 \end{pmatrix},$$

where we use dashed lines to represent the partition leading to $N_5^{(0)} = [0]$, $N_5^{(1)} = [1, 2]$ and $N_5^{(2)} = \mathbf{0}_{2 \times 2}$.

The Introduce node $i = 6$ introduces vertex $v = 5$ and simply adds an additional type II row and column, indexed by 5, filled with zeros to produce matrix

$$N_6 = \begin{matrix} 2 \\ 1 \\ 4 \\ 5 \end{matrix} \begin{pmatrix} 0\ 1\ 2\ 0 \\ 1\ 0\ 0\ 0 \\ 2\ 0\ 0\ 0 \\ 0\ 0\ 0\ 0 \end{pmatrix}.$$

Node $i = 7$ is of type Join. We apply JoinBox with input matrices N_3 and N_6. This creates a matrix N_7^* as in (6.3), where the rows of type I coming from the different branches are stacked on top of each other and the matrices corresponding to rows of type II are added to each other. Since N_3 has no type I rows, this leads to

$$N_7 = \begin{matrix} 2 \\ 1 \\ 4 \\ 5 \end{matrix} \begin{pmatrix} 0\ \ \ 1\ \ \ 2\ 0 \\ 1\ -1\ \ \ 1\ 0 \\ 2\ \ \ 1\ -1\ 0 \\ 0\ \ \ 0\ \ \ 0\ 0 \end{pmatrix},$$

which satisfies all the required properties and is ready to be transmitted to the node's parent.

Node $i = 8$ forgets vertex $v = 4$. The first step is to update $N_7^{(2)}$ by adding the entries of M to the row and column associated with vertex $v = 4$, which gives

$$N_8^{*(2)} = \begin{matrix} 1 \\ 4 \\ 5 \end{matrix} \begin{pmatrix} -1\ 1\ 0 \\ \ \ 1\ 0\ 0 \\ \ \ 0\ 0\ 0 \end{pmatrix}.$$

Exchanging rows and columns of N_8^* so that the first row is indexed by $v = 4$, we turn the matrix into the format of Eq. (6.6):

$$N_8^* = \begin{matrix} 4 \\ 2 \\ 1 \\ 5 \end{matrix} \begin{pmatrix} 0 & 2 & 1 & 0 \\ 2 & 0 & 1 & 0 \\ 1 & 1 & -1 & 0 \\ 0 & 0 & 0 & 0 \end{pmatrix}.$$

We are in case 2 of ForgetBox with $d_v = 0$, $\mathbf{x}_v = [2]$, $\mathbf{y}_v = [1, 0]$. Since $\mathbf{x}_v = [2]$ has a single nonzero entry, no operation is performed to annihilate elements of $\mathbf{x}_v$.

To transform the left upper corner of the matrix $\begin{pmatrix} 0 & 2 \\ 2 & 0 \end{pmatrix}$ into a diagonal matrix,

we perform the operations in (6.11), namely $R_2 \leftarrow R_2 + \frac{1}{2}R_4$, $C_2 \leftarrow C_2 + \frac{1}{2}C_4$ followed by $R_4 \leftarrow R_4 - R_2$ and $C_4 \leftarrow C_4 - C_2$. This leads to the matrix

$$A_2 = \begin{matrix} 4 \\ 2 \\ 1 \\ 5 \end{matrix} \begin{pmatrix} -2 & 0 & -1/2 & 0 \\ 0 & 2 & 3/2 & 0 \\ -1/2 & 3/2 & -1 & 0 \\ 0 & 0 & 0 & 0 \end{pmatrix}.$$

We now use the nonzero pivots obtained in order to diagonalize R_4 and R_2. To achieve this, we perform the operations $R_1 \leftarrow R_1 - \frac{1}{4}R_4$, $C_1 \leftarrow C_1 - \frac{1}{4}C_4$, followed by $R_1 \leftarrow R_1 - \frac{3}{4}R_2$, $C_1 \leftarrow C_1 - \frac{3}{4}C_2$. This produces the matrix

$$A_3 = \begin{matrix} 4 \\ 2 \\ 1 \\ 5 \end{matrix} \begin{pmatrix} -2 & 0 & 0 & 0 \\ 0 & 2 & 0 & 0 \\ 0 & 0 & -2 & 0 \\ 0 & 0 & 0 & 0 \end{pmatrix}.$$

At this point, the first two rows are diagonalized, corresponding to vertices 4 and 2. This means that the pairs $(4, -2)$ and $(2, 2)$ are appended to the global array D and that node $i = 8$ transmits the matrix

$$N_8 = N_8^{(2)} = \begin{matrix} 1 \\ 5 \end{matrix} \begin{pmatrix} -2 & 0 \\ 0 & 0 \end{pmatrix},$$

so that $N_8^{(0)}$ and $N_8^{(1)}$ are empty. Its parent $i = 9$ forgets vertex $v = 5$. We first update the entries in the row and column associated with $v = 5$ using entries of M. Doing this and exchanging rows and columns lead to

$$N_9^* = \begin{matrix} 5 \\ 1 \end{matrix} \begin{pmatrix} -1 & -1 \\ -1 & -2 \end{pmatrix},$$

where $d_v = -1 \neq 0$, x_v is empty, and $y_v = [-1]$. ForgetBox directs us to use d_v to diagonalize the row/column corresponding to v, so that we perform $R_1 \leftarrow R_1 - R_5$ and $C_1 \leftarrow C_1 - C_5$, which produces

$$\begin{matrix} 5 \\ 1 \end{matrix} \begin{pmatrix} -1 & 0 \\ 0 & -1 \end{pmatrix}.$$

The pair $(v, d_v) = (5, -1)$ is added to v, and the box $N_9 = [-1]$ is transmitted to its parent, with a single row of type II, indexed by vertex $v = 1$. Node $i = 10$ forgets vertex $v = 1$. Actually, it simply updates N_9 to $[d_1]$, where $d_1 = -1 + m_{11}$, and adds $(1, d_1)$ to D, as x_v and y_v are both empty in this case. We have $d_1 = 0$, which means that the global array returned by the algorithm is

$$\begin{pmatrix} v \\ d_v \end{pmatrix} = \begin{pmatrix} 3 & 4 & 2 & 5 & 1 \\ 1 & -2 & 2 & -1 & 0 \end{pmatrix}.$$

This means that M is congruent to the diagonal matrix

$$D = \begin{pmatrix} 0 & 0 & 0 & 0 & 0 \\ 0 & 2 & 0 & 0 & 0 \\ 0 & 0 & 1 & 0 & 0 \\ 0 & 0 & 0 & -2 & 0 \\ 0 & 0 & 0 & 0 & -1 \end{pmatrix}.$$

We also conclude that M has 2 positive eigenvalues, 2 negative eigenvalues, and 0 is an eigenvalue with multiplicity 1.

To conclude this chapter, we offer some directions for further investigation. As discussed in Chap. 4, the main application for graph widths has been the design of efficient algorithms for NP-complete or even harder problems for graphs whose width is bounded above by an absolute constant. This is known as *fixed parameter tractability* . However, the algorithm in this chapter dealt with a problem that may be solved in cubic time with a standard approach using elementary row and column operations. The main contribution of this algorithm was to achieve linear time for matrices associated with graphs with more structure, namely for graphs with bounded treewidth. This fits into the trend of "FPT within P" (see [4]), which investigates fundamental problems that are solvable in polynomial time, but for which a lower exponent may be achieved using an FPT algorithm that is also polynomial in terms of the parameter k.

There are many open problems in this direction. For instance, it is not known whether any $n \times n$ matrix whose underlying graph has treewidth at most k may be diagonalized in time $O(k^2 n)$ using Gaussian elimination,[6] the best currently

---

[6] This would be best possible up to the constant implicit in the O notation.

known algorithm [1] runs in time $O(k^3 n)$. Recall that Gaussian elimination only allows elementary row operations. In this chapter, we have shown how to do this for *symmetric* matrices if we are allowed to perform column operations, which was crucial to deal with accidental cancelations.

We also believe that the algorithm in this chapter will be a valuable theoretical tool to derive results about spectral parameters in graph theory, as has already been the case for the simpler algorithms of Chaps. 3 and 5. This includes bounds on the index, the spectral radius, and other eigenvalues with important roles in applications, such as the algebraic connectivity. These algorithms have proven to be particularly useful for bounding parameters that depend on the entire spectrum, such as most graph energies, and to obtain results about eigenvalue multiplicity. References may be found in Chaps. 3 and 5. An important step in these applications is to understand how the graph structure affects the algorithm, and this may not be easy to do for general graphs. We think that looking for results in graph classes with small treewidth and well-understood tree decompositions would be the natural first step to achieve more general results of this type.

References

1. Fomin, F.V., Lokshtanov, D., Saurabh, S., Pilipczuk, M., Wrochna, M.: Fully polynomial-time parameterized computations for graphs and matrices of low treewidth. ACM Trans. Algorithms **14**(3), 34:1–34:45 (2018)
2. Fürer, M., Hoppen, C., Trevisan, V.: Efficient Diagonalization of Symmetric Matrices Associated with Graphs of Small Treewidth. In: A. Czumaj, A. Dawar, E. Merelli (eds.) 47th International Colloquium on Automata, Languages, and Programming (ICALP 2020), Leibniz International Proceedings in Informatics (LIPIcs), vol. 168, pp. 52:1–52:18. Schloss Dagstuhl–Leibniz-Zentrum für Informatik, Dagstuhl, Germany (2020). https://doi.org/10.4230/LIPIcs. ICALP.2020.52. https://drops.dagstuhl.de/opus/volltexte/2020/12459
3. Fürer, M., Hoppen, C., Trevisan, V.: Efficient diagonalization of symmetric matrices associated with graphs of small treewidth (2021)
4. Giannopoulou, A.C., Mertzios, G.B., Niedermeier, R.: Polynomial fixed-parameter algorithms: A case study for longest path on interval graphs. Theor. Comput. Sci. **689**, 67–95 (2017)
5. Radhakrishnan, V., Hunt, H.B., Stearns, R.E.: Efficient algorithms for solving systems of linear equations and path problems. In: A. Finkel, M. Jantzen (eds.) STACS 92, pp. 109–119. Springer, Berlin, Heidelberg (1992)
6. Rose, D.J.: A graph-theoretic study of the numerical solution of sparse positive semidefinite systems of linear equations. In: R.C. READ (ed.) Graph Theory and Computing, pp. 183–217. Academic Press (1972). https://doi.org/10.1016/B978-1-4832-3187-7.50018-0. https://www. sciencedirect.com/science/article/pii/B9781483231877500180

Chapter 7
Locating Eigenvalues Using Slick Clique Decomposition

7.1 Clique-Width and Diagonalization

In Chap. 5, we presented a diagonalization algorithm for the adjacency matrix of a cograph G. It runs in linear time and performs congruence operations, using the cotree representation of G. It works bottom-up on the cotree, and at each stage, it diagonalizes the rows and columns associated with either one or two vertices, which are siblings in the terminology of Chap. 4 and then removes the corresponding leaves from the tree. Here, we describe a generalization of this approach to arbitrary graphs, introduced in [4, 5], using a parse tree representation of a clique decomposition (see Sects. 4.5 and 4.6). This leads to an $O(k^2n)$ time diagonalization algorithm for the adjacency matrix of graphs having clique-width at most k. We notice that such matrices often have $\Omega(n^2)$ nonzero entries and that the clique-width may be a small constant even if other parameters, such as the treewidth, are linear in n. However, the algorithm in this chapter is less general than the algorithm in Chap. 6, as it seems impossible to extend it to arbitrary symmetric matrices. On the other hand, it can be extended in a straightforward way to any symmetric matrix M whose underlying graph has clique-width at most k and whose nonzero off-diagonal entries are all equal to each other. This includes many widely used graph matrices, such as the adjacency, the Laplacian, and the signless Laplacian matrices.

The algorithm `Diagonalize Cograph` of Chap. 5 exploited the fact that in any cograph of order $n \geq 2$, there exist two vertices u and v for which either $N(u) = N(v)$ or $N[u] = N[v]$. This means that their corresponding rows and columns in the adjacency matrix can differ by at most two positions. By subtracting, say, the row (column) of u from the row (column) of v, the off-diagonal entries of the row and column of v are annihilated except possibly one such entry. The following analog is crucial to our algorithm and uses the binary parse tree described in Sect. 4.6.

Remark 7.1 Let T_G be a parse tree for a slick expression of a graph $G = (V, E)$ with adjacency matrix A, and let Q be a node in T_G. If two vertices u and v have the same label at Q, then their rows (columns) are the same outside of the matrix for the subtree rooted at Q, that is, if $w \in V$ is not associated with a leaf of this subtree, then $A_{u,w} = A_{v,w} = A_{w,u} = A_{w,v}$.

To see why Remark 7.1 is true, let P be the lowest common ancestor of Q and w. Consider the path from Q to P. While the labels of u and v can change along this path, their labels must be equal. The adjacencies are determined by the label at P. Therefore, u is adjacent to w if and only if v is adjacent to w.

In spite of this similarity, the new algorithm requires several new ingredients. Indeed, unlike the cograph algorithm, the new algorithm does not diagonalize a given number of vertices at each node of the parse tree. Instead, as was done in the algorithm `Diagonalize Treewidth` of Chap. 6, it transmits information up the tree in a very compact way, also using a *box*.

Another property that is crucial in the cograph algorithm is that subgraphs generated by subexpressions are induced subgraphs. Unfortunately, this does not hold for the k-expressions of Sect. 4.5, where an edge creating operation applied after a join of G and H typically introduces edges within G and within H. This is exactly the purpose of using the slick clique decomposition of Sect. 4.6, for which this property holds by Proposition 4.3(b). As pointed out in Theorem 4.6, there are linear time transformations to translate a k-expression into a slick k-expression and a slick k-expression into a $2k$-expression. Hence, we may assume that we are given either a clique decomposition or a slick clique decomposition of width k for the graph G with n vertices.

7.2   The Algorithm

We now describe the diagonalization algorithm. Let $G = (V, E)$ be a graph on $n = |V|$ vertices with adjacency matrix A, and assume that it is generated by a slick expression Q_G.

Given a real number x, we find a diagonal matrix congruent to $B = A + xI_n$. The expression Q_G is associated with a parse tree T_G having $2n - 1$ nodes. As defined in Sect. 4.6, this is a rooted binary tree where the n leaves are labeled by the operators $i(v)$ and the internal nodes contain operations of type $\oplus_{S,R,L}$. Additionally, the left child corresponds to the root of the left subgraph, while the right child corresponds to the root of the right subgraph. As in the previous chapters, the algorithm `Diagonalize Clique-width` works bottom-up in the parse tree T_G. In fact, it does a postorder traversal and operates on Q after its children have been processed. Figure 7.1 gives a high-level description of the algorithm.

```
input: the parse tree T of slick k-expression Q_G for G, a scalar x
output: diagonal matrix D = diag(d_1, ..., d_n) congruent to A(G) + xI

Diagonalize Clique-width(G,x)
    Order the vertices of T as Q_1, Q_2, ..., Q_{2n-1} = Q_G in postorder
    for t from 1 to 2n - 1 do
        if is-leaf(Q_t) then b_{Q_t}=LeafBox(Q_t, x)
        else if Q_t = Q_ℓ ⊕_{S,L,R} Q_r then b_{Q_t} = CombineBoxes(b_{Q_ℓ}, b_{Q_r})
    end for
    DiagonalizeBox(b_{Q_{2n-1}})
```

Fig. 7.1 High-level description of the algorithm `Diagonalize Clique-width`

As was the case for algorithm `Diagonalize Treewidth` in Chap. 6, a node Q in the tree produces a data structure that we call a *k-box* b_Q. Here this is a 4-tuple $[k', k'', M, \Lambda]$, where k' and k'' are non-negative integers bounded above by k, M is an $m \times m$ symmetric matrix, where $m = k' + k'' \leq 2k$, and Λ is a vector whose m entries are labeled in $\{1, \ldots, k\}$. The algorithm traverses the parse tree from the leaves to the root so that, at each node, the algorithm either initializes a box or combines the boxes produced by the node's two children into its own box, transmitting it to its parent. At each node, the algorithm operates on a small $O(k) \times O(k)$ matrix, performing congruence operations. These operations represent operations that would be performed on the large $n \times n$ matrix B in order to diagonalize it, and this small matrix represents a *partial view* of the actual $n \times n$ matrix. While processing the node, the algorithm may also produce diagonal elements of a matrix congruent to $A + xI_n$. These diagonal elements are appended to a global array as they are produced.

We shall now describe each of these stages in detail. Recall that the subtree at any node Q_t is associated with a subexpression. Let n_{Q_t} be the order of the induced subgraph $G(Q_t)$ of G generated by this subexpression, and let A_{Q_t} be its adjacency matrix. Let I_{Q_t} be the $n_{Q_t} \times n_{Q_t}$ identity matrix, and $B_{Q_t} = A_{Q_t} + xI_{Q_t}$.

The goal at each node Q_t is to construct, by means of congruence operations, a matrix $\widetilde{B}_{Q_t}$ associated with $G(Q_t)$ that is diagonal except for at most $2k$ rows and columns having the form of Eq. (7.1). The diagonal elements $d_1, \ldots, d_\ell$ may have been computed in earlier stages.

$$
\widetilde{B}_{Q_t} = \begin{pmatrix} d_1 & & & 0 \\ & \ddots & & \\ & & d_\ell & \\ 0 & & & M_{Q_t} \end{pmatrix}.
\tag{7.1}
$$

The partition of the matrix M_{Q_t} is given by the box b_{Q_t} in Eq. (7.2). As in Chap. 6, $M_t^{(0)} = \mathbf{0}_{k' \times k'}$, $M_t^{(2)}$ is a $k'' \times k''$ symmetric matrix, and $M_t^{(1)}$ is a $k' \times k''$ matrix in row echelon form with pivots in every row:

$$
M_{Q_t} = \begin{array}{|c|c|}
\hline
M_t^{(0)} & M_t^{(1)} \\
\hline
M_t^{(1)T} & M_t^{(2)} \\
\hline
\end{array}
\tag{7.2}
$$

Note that k' can be zero in which case we regard $M_t^{(0)}$ and $M_t^{(1)}$ as empty.

Equation (7.3) below illustrates how the matrix $\widetilde{B}_{Q_t}$ of Eq. (7.2) fits into $\widetilde{B}_{Q_t}^+$. The reason why the algorithm works is that $\widetilde{B}_{Q_t}$ is congruent to b_{Q_t} at the end of each step t, while being a submatrix of a matrix $\widetilde{B}_{Q_t}^+$ that is congruent to the input matrix B. For clarity, we assume that the rows and columns corresponding to vertices that have already been diagonalized or lie in $G(Q_t)$ appear first. We emphasize that $\widetilde{B}_{Q_t}$ and $\widetilde{B}_{Q_t}^+$ are not explicitly computed by the algorithm.

$$
\widetilde{B}_{Q_t}^+ = \left(
\begin{array}{ccc}
D & 0 & 0 \\
& \begin{array}{|c|c|}
\hline
M^{(0)} & M^{(1)} \\
\hline
M^{(1)T} & M^{(2)} \\
\hline
\end{array}
&
\begin{array}{ccc}
0 & \cdots & 0 \\
\vdots & & \vdots \\
0 & \cdots & 0 \\
\beta_1^1 & \cdots & \beta_1^s \\
\vdots & & \vdots \\
\beta_{k''}^1 & \cdots & \beta_{k''}^s
\end{array}
\\[4ex]
\begin{array}{c}
0 \\
\vdots \\
0
\end{array}
&
\begin{array}{ccccc}
0 & \cdots & 0\,\beta_1^1 & \cdots & \beta_{k''}^1 \\
\vdots & & \vdots\,\vdots & & \vdots \\
0 & \cdots & 0\,\beta_1^s & \cdots & \beta_{k''}^s
\end{array}
& M'
\end{array}
\right) .
\tag{7.3}
$$

The right side of (7.3) shows how the remainder of the large matrix B is transformed as a side effect of operating on the small submatrix M_{Q_t}. The diagonal matrix D represents *all* diagonalized elements produced up to step t in the algorithm. Also the k' rows and columns of $M^{(0)}$ extend with zero vectors, defining the boundary between $M^{(0)}$ and $M^{(2)}$.

The β_i^j are zero-one entries in the partially diagonalized matrix, whose relation with the corresponding entries in the original matrix B will be explained in Lemma 7.1 below. It is important to keep in mind that after node Q_t has been processed, all vertices in the subgraph $G(Q_t)$ correspond to rows in D or M_{Q_t}. Some rows of D may correspond to vertices outside of $G(Q_t)$, which have been diagonalized in earlier steps. The submatrix M' in (7.3) contains all undiagonalized rows $w \notin G(Q_t)$, may include rows in $M_{Q_{t'}}$ for $t' \neq t$, and is empty after the last iteration of the algorithm.

As in Chap. 6, we say that the k' rows in $M^{(0)}$ have *type I*, and that the k'' rows of $M^{(2)}$ have *type II*. Moreover, these rows are indexed by the rows of the original matrix, and each of them has a label in $[k]$ coming from the slick expression that defines $G(Q_t)$. An advantage here is that each row of the original matrix is associated with a single leaf of the parse tree, so that the sets of type I and type II rows coming from different branches of the tree are always disjoint. The important information is $M = M_{Q_t}$, these labels, and the integers k' and k'', which are all stored in the k-box (or simply box) $b_{Q_t} = [k', k'', M, \Lambda]$ mentioned above, whose *size* is $k' + k''$. To ensure that $M^{(2)}$, whose rows have type II, has order $k'' \leq k$ at the end of step t, each label appears in at most one type II row. In the algorithm of this chapter, a row always begins as a type II row, then becomes a type I row, and finally becomes diagonalized.

When the node Q_t is a leaf corresponding to a subexpression $i(v)$, $B_{Q_t} = [x]$, where x is the value as in Fig. 7.1. This means that the box contains a 1×1 matrix $M_{Q_t} = [x]$, whose row is labeled i, $k' = 0$, and $k'' = 1$. The procedure LeafBox is illustrated in Fig. 7.2.

Next we will show how CombineBoxes works, that is, we explain how an internal node produces its box from the boxes transmitted by its children. Before describing this process, we state a lemma that summarizes facts about the algorithm and is needed to establish its correctness. For simplicity, when referring to operations, we always mention the row operations, with the understanding that the corresponding column operations are also performed. We also identify vertices with their rows and columns. Informally, part (a) states that diagonal elements must be produced by CombineBoxes and that rows are not modified after being diagonalized. Part (b) ensures that if v has type I in M_{Q_t}, the entries of its row in $\widetilde{B}_{Q_t}^+$ corresponding to vertices outside M_{Q_t} are all 0, as depicted in (7.3). Part (c)

```
LeafBox(Q,x)
    input: k-expression Q = i(v), a scalar x
    output: [0, 1, [x], [i]]
```

Fig. 7.2 Procedure LeafBox

tells us that no entry uv can be modified by the algorithm, while both vertices have type II. Part (d) tells us that the algorithm cannot modify entry uv at step t if neither u nor v corresponds to a row in the box. Part (e) tells us that a type I vertex cannot turn back to type II.

Lemma 7.1 *Let $Q_1, Q_2, \ldots, Q_{2n-1} = Q_G$ be the nodes of the parse tree of the slick k-expression Q_G that defines $G = (V, E)$, listed in postorder. Consider the matrices B, M_{Q_τ}, $\widetilde{B}_{Q_\tau}$, and $B_{Q_\tau}^+$ defined above, where $\tau \in \{1, \ldots, 2n-1\}$, and let $v, w \in V$.*

(a) *If row v is diagonalized at step τ, then step τ applied CombineBoxes. Moreover, if row v was diagonalized when processing $Q_{\tau'}$ for some $1 \leq \tau' < \tau$, then no entries in this row are modified in step τ.*

(b) *Suppose that row v has type I in $M_{\tau'}$, for some $\tau' \in \{1, \ldots, \tau\}$. Suppose also that w is a row such that, for all $j \in \{\tau', \ldots, \tau\}$, v and w are not simultaneously in $G(Q_j)$. Then the entry vw in $B_{Q_\tau}^+$ is 0.*

(c) *Suppose that row v has type II in M_{Q_τ} and $w \notin G(Q_\tau)$. Suppose also that, for all $\tau' < \tau$, w has type II in $\widetilde{B}_{Q_{\tau'}}$ whenever $w \in G(Q_{\tau'})$. Then $B_{Q_\tau}^+$ and B are equal in position vw.*

(d) *Suppose that, for some $\tau' \in \{0, \ldots, \tau - 1\}$, the vertices v and w are both not in $G(Q_j)$ for all $\tau' < j \leq \tau$. Then $B_{Q_\tau}^+$ and $B_{Q_{\tau'}}^+$ are equal in position vw (where $B_{Q_0}^+$ is defined to be B).*

(e) *Suppose that row v has type I in $M_{Q_{\tau'}}$, for some $\tau' \in \{1, \ldots, \tau - 1\}$. For all $\tau' < j \leq \tau$, if v is in M_{Q_j}, then it has type I.*

Lemma 7.1 may be proved by induction on τ. The statement is trivial for $\tau = 1$, and we refer the reader to [5, Lemma 4] for a detailed proof.

Suppose that Q_t is a node with children Q_ℓ and Q_r, that is $Q_t = Q_\ell \oplus_{S,L,R} Q_r$. Let M_{Q_ℓ} and M_{Q_r} denote the matrices in the boxes transmitted, respectively, by Q_ℓ and Q_r, containing the undiagonalized rows in $\widetilde{B}_{Q_\ell}$ and $\widetilde{B}_{Q_r}$. The goal of Q_t is to combine M_{Q_ℓ} and M_{Q_r} into a single matrix of size at most $2k$ representing $\widetilde{B}_{Q_t}$. More precisely, based on M_{Q_ℓ} and M_{Q_r}, node Q_t first constructs the submatrix M of $B_{Q_{t-1}}^+$ formed by the undiagonalized rows in $\widetilde{B}_{Q_\ell}$ and $\widetilde{B}_{Q_r}$. Recall that this cannot be done directly, as $B_{Q_{t-1}}^+$ is never actually computed by the algorithm. The following result tells us how this can be done.

Lemma 7.2 *Suppose that $Q_t = Q_\ell \oplus_{S,L,R} Q_r$ and consider the submatrix M of $B_{Q_{t-1}}^+$ formed by the undiagonalized rows in $\widetilde{B}_{Q_\ell}$ and $\widetilde{B}_{Q_r}$. The following hold:*

(a) *If $s \in \{\ell, r\}$ and $v, w \in G(Q_s)$, then the entry vw in M is equal to the entry vw in M_{Q_s}.*

(b) If $v \in G(Q_\ell)$, $w \in G(Q_r)$, and at least one of v and w has type I, then the entry vw in M is 0.

(c) If $v \in G(Q_\ell)$, $w \in G(Q_r)$, and both have type II, then the entry vw in M is 1 if $(i, j) \in S$, where i is the label of v and j is the label of w; otherwise, it is 0.

Proof Consider the entry vw in a submatrix M as in the statement of the lemma. By the definition of slick expression, $V(Q_\ell) \cap V(Q_r) = \emptyset$.

For part (a), assume first that both v and w are on the same side, say $v, w \in G(Q_\ell)$. Because the nodes of the parse tree are processed in postorder, any Q_j, for $\ell < j < t$, cannot be in the same branch of Q_ℓ. Hence, v and w cannot be in $G(Q_j)$ for $\ell < j < t$. We observe that from the postorder, we have $\ell < t - 1$. Now, Lemma 7.1(d) applied to $\tau' = \ell$ and $\tau = t - 1$ implies that the entry vw is the same in $\widetilde{B}^+_{Q_{t-1}}$ and $\widetilde{B}^+_{Q_\ell}$. By the definition of M_{Q_ℓ} in (7.1), the entries vw in $\widetilde{B}^+_{Q_\ell}$ and M_{Q_ℓ} are equal. The desired conclusion follows because M is a submatrix of $\widetilde{B}^+_{Q_{t-1}}$. When $v, w \in G(Q_r)$, the postorder implies that $r = t - 1$ and trivially holds that the entry vw is the same in $\widetilde{B}^+_{Q_{t-1}}$ and $\widetilde{B}^+_{Q_r}$, and the same conclusion holds.

For part (b), fix $v \in G(Q_\ell)$ and $w \in G(Q_r)$. First assume that the row of v has type I in M_{Q_ℓ}. Note that $w \notin G(Q_\ell)$ and v is not in $G(Q_j)$ for $j \in \{\ell+1, \ldots, t-1\}$. By Lemma 7.1(b) for $\tau' = \ell$ and $\tau = t - 1$, the entry vw in $\widetilde{B}^+_{Q_{t-1}}$ is equal to 0. The same conclusion would be achieved if the row of w has type I in M_{Q_r}.

We finally consider the case where the rows of v and w have type II in M_{Q_ℓ} and M_{Q_r}, respectively. By postorder, we know that $\ell < r = t - 1$. We know that row v has type II in M_{Q_ℓ} and that $v \notin G(Q_j)$ for all $j \in \{\ell + 1, \ldots, t - 1\}$. Moreover, Lemma 7.1(e) implies that, if $j \in \{1, \ldots, \ell - 1\}$ and $v \in G(Q_j)$, then row v has type II. We may apply Lemma 7.1(c) for $\tau = r = t - 1$ and $\tau' = \ell$ to conclude that $\widetilde{B}^+_{Q_r} = \widetilde{B}^+_{Q_{t-1}}$ and B are equal in position vw. To conclude the proof, since $Q_t = Q_\ell \oplus_{S,L,R} Q_r$, we know that the entry vw in B is 1 if $(i, j) \in S$, where i is the label of v and j is the label of w; otherwise, it is 0. The desired conclusion follows because, by induction, the entry vw of M equals the entry vw in $\widetilde{B}^+_{Q_{t-1}}$. $\square$

Lemma 7.2 implies that, when processing Q_t, the matrix M may be constructed by first taking the disjoint union of the matrices transmitted by its children and then updating the entries vw, where v and w are type II vertices of different sides. Precisely, if $(i, j) \in S$, v is a type ii vertex in $G(Q_\ell)$ with label i, and w is a type II vertex in $G(Q_r)$ with label j, we place a one in the row (column) of v and column (row) of w. Observe that the unique label condition imposed on $M^{(2)}$ implies that at most one pair of entries will be modified for any element of S. Let F be the block of ones defining these edges. Then node Q_t starts with a matrix as in equation

$$
M_{Q_t} =
\begin{pmatrix}
\begin{array}{cc}
0 & \begin{matrix} * & * & * \\ & * & * \\ & & * \end{matrix}
\end{array}
& 0 & 0 \\[2mm]
\begin{array}{cc}
\begin{matrix} * & \\ * & * \\ * & * & * \end{matrix} & M_\ell^{(2)}
\end{array}
& 0 & F \\[4mm]
0 \qquad 0 & 0 \; \begin{matrix} * & * & * & * \\ & * & * & * \\ & & * & * \\ & & & * \end{matrix} \\[4mm]
0 \qquad F^T & \begin{matrix} * & \\ * & * \\ * & * & * \\ * & * & * & * \end{matrix} \; M_r^{(2)}
\end{pmatrix}.
\tag{7.4}
$$

Here, and in the remainder of the description of the procedure CombineBoxes, we abuse the notation slightly and use M_{Q_t} to refer to the matrix obtained by merging the boxes transmitted by the children, even before it is a proper matrix for the box b_{Q_t}. Next Q_t relabels the rows of M_{Q_ℓ} and M_{Q_r}, using the functions L and R, respectively.

Using permutations of rows and columns, we combine the k'_ℓ type I rows from M_{Q_ℓ} with the k'_r type I rows of M_{Q_r} and combine the k''_ℓ type II rows in M_{Q_ℓ} with the k''_r type II rows in M_{Q_r}, so that they are contiguous in M_{Q_t}.

We observe that the zero pattern outside the matrices M_{Q_ℓ} and M_{Q_r} implies that the type I rows from M_{Q_ℓ} and M_{Q_r} are still type I in (7.4). Additionally, from the fact that the type I rows of M_{Q_ℓ} and M_{Q_r} are distinct, we see that the new $M^{(1)}$ of M_{Q_t}, which is formed by placing $M_\ell^{(1)}$ on top of $M_r^{(1)}$, is already in row echelon form. We illustrate in Eq. (7.5) the transformation of the matrix in (7.4) after we performed the permutations described above. Hence, these permutations of rows and columns lead to a matrix M_Q in the form of (7.2). In this matrix, $k' = k'_\ell + k'_r$ and $k'' = k''_\ell + k''_r$. By construction, we have $k' \leq k''$; however, we are not guaranteed that $k'' \leq k$.

$$M_{Q_t} = \begin{bmatrix} 0 & 0 & \begin{matrix} * & * & * \\ * & * & \\ * & & \end{matrix} & 0 \\ 0 & 0 & 0 & \begin{matrix} * & * & * & * \\ * & * & * & \\ * & * & & \\ * & & & \end{matrix} \\ \begin{matrix} * & & \\ * & * & \\ * & * & * \end{matrix} & 0 & M_\ell^{(2)} & F \\ 0 & \begin{matrix} * & & & \\ * & * & & \\ * & * & * & \\ * & * & * & * \end{matrix} & F^T & M_r^{(2)} \end{bmatrix}. \tag{7.5}$$

This requires achieving $k'' \le k$ without losing $k' \le k''$ using congruence operations. As a byproduct, new permanent elements may be added to the diagonal submatrix D of $\widetilde{B_{Q_t}^+}$. The matrix M_{Q_t} is shrunk in two steps:

(i) If two type II rows j and j' have the same label, then congruence operations are performed to turn one of them, say $v = j'$, into type I.

(ii) Congruence operations are performed with the aim of diagonalizing the type I row corresponding to v, or of inserting v into M_{Q_t} in a way that $M_{Q_t}^{(0)}$ is the zero matrix and $M_{Q_t}^{(1)}$ is in row echelon form and has a pivot on each row.

Note that if steps (i) and (ii) are repeatedly applied to pairs of type II rows with the same label until there are no such pairs, we immediately obtain $k'' \le k$. By ensuring that $M_{Q_t}^{(1)}$ remains in row echelon form with pivots on each row (or that $M_{Q_t}^{(1)}$ becomes empty), we ensure that $k' \le k''$.

To achieve (i), fix that two type II rows j and j' have the same label. By Remark 7.1, their rows and columns must agree outside of M_{Q_t}. Therefore, in (7.3), $\beta_j^i = \beta_{j'}^i$ for all i. Performing the operations

$$R_{j'} \leftarrow R_{j'} - R_j, \; C_{j'} \leftarrow C_{j'} - C_j \tag{7.6}$$

eliminates any nonzero entries of row j' in $\widetilde{B_{Q_t}^+}$, except possibly for the columns within M_{Q_t}, and thus transforms the type II row j' into a type I row.

Part (ii) is done in the same fashion as procedure ForgetBox of Chap. 6. We recall some of the reasoning here. Suppose that v is the row to be inserted. For convenience, after exchanging rows and columns, we look at M_{Q_t} in Eq. (7.7).

Here, the first row and column represent the row and column in M_{Q_t} associated with v. In particular, d_v is the diagonal element, while x_v is the vector whose elements correspond to the other type I rows in $M_{Q_t}^{(0)}$. The vector y_v represents the correspondence with the type II rows in $M_{Q_t}^{(2)}$. We observe that the lengths of x_v and y_v are k' and $k'' - 1$, respectively.

$$M_{Q_t} = \begin{array}{|c|c|c|} \hline d_v & \mathbf{x}_v & \mathbf{y}_v \\ \hline \mathbf{x}_v^T & \mathbf{0}_{k' \times k'} & M_{Q_t}^{(1)} \\ \hline \mathbf{y}_v^T & M_{Q_t}^{(1)T} & M_{Q_t}^{(2)} \\ \hline \end{array}. \tag{7.7}$$

Depending on the value of d_v, and the vectors $\mathbf{x}_v$ and $\mathbf{y}_v$, we proceed in different ways. The preferred outcome is to diagonalize the row/column associated with v, and this is particularly easy when $d_v \neq 0$. Indeed, in this case, we simply use d_v to annihilate all the elements in row/column v, performing the congruence operations. Note that due to the zero extension of the rows (see Lemma 7.1(b)), when the congruence operations are performed to produce $\widetilde{B}_{Q_t}^+$, all the work is restricted to the small matrix M_{Q_t}, and the *shape* of $\widetilde{B}_{Q_t}^+$ does not change.

In general, we deal with v exactly as in procedure ForgetBox of Fig. 6.5 in Chap. 6. We refer the reader to Fig. 6.5 and to the discussion following it for the full description of each case and its motivation. We recall that the row corresponding to v and an additional type I row are diagonalized with diagonal entries of opposite signs if $\mathbf{x}_v \neq 0$ (case 2 of CombineBoxes), v is diagonalized with diagonal element d_v if $\mathbf{x}_v$ is empty or zero and $d_v \neq 0$ (subcases 1a and 1c), or if $d_v = 0$ and $\mathbf{x}_v$ and $\mathbf{y}_v$ are both empty or zero (subcase 1a). The only remaining case is when $d_v = 0$, $\mathbf{x}_v$ is empty or zero, and $\mathbf{y}_v \neq 0$. Then the algorithm performs steps to insert $\mathbf{y}_v$ into the matrix in row echelon form (subcase 1b of CombineBoxes).

This concludes the description of the procedure CombineBoxes, which appears in Fig. 7.3. Note that all operations performed in the procedure are congruence operations. To conclude the description of the algorithm, we can assume that at the root

$$Q_G = Q_\ell \oplus_{L,R,S} Q_r,$$

L and R map all vertices to the same label, as labels are no longer needed. After applying CombineBoxes, we will obtain $k'' = 1$, and $M^{(0)}$ is either zero or empty. If it is empty, then the 1×1 matrix $M^{(2)}$ contains the final diagonal element. Otherwise, M_Q is a 2×2 matrix having form

$$\begin{pmatrix} 0 & a \\ a & b \end{pmatrix}.$$

If $b \neq 0$, it can be made fully diagonal using the congruence operations in (6.8) (where u and v denote the first and second rows, respectively). If $b = 0$, it can be

input: two k-boxes b_{Q_ℓ} and b_{Q_r}
output: a k-box b_Q

```
CombineBoxes(b_{Q_ℓ},b_{Q_r})
    form matrix M as in (7.4);
    relabel rows with functions L and R;
    combine type-i rows (columns), combine type-ii rows (columns);
    ensure type-ii rows have distinct labels:
    for each pair (v, w) of type-ii rows with equal labels
        perform operations as in (7.6);
        Insert new type-i row:
            if x_v is empty or 0 then
                if y_v is empty or 0 then        //subcase 1a
                    add (v, d_v) to D
                    remove row v from M
                else if d_v = 0 then             //subcase 1b
                    do row/column operations as in (6.7)
                    if row v gets a pivot then insert the row into M^(1)
                    else add 0 to D and remove row from M
                else // Here {d_v ≠ 0, y_v ≠ 0}.  //subcase 1c
                    use d_v to diagonalize row/column v as in (6.8)
                    add d_v do D and remove row v form M
            else // Here x_v ≠ 0.    //case 2
                let u be the vertex of the rightmost nonzero entry of x_v
                if x_v has other nonzero entries then
                    eliminate them with the operations (6.9)
                if d_v ≠ 0 then perform the operations (6.10)
                perform the operations (6.11)
                use d_v and d_u to diagonalize rows/columns v and u as in (6.7)
                add d_v and d_u to D and eliminate rows v and u from M.
```

Fig. 7.3 Procedure CombineBoxes

made fully diagonal using the congruence operations in (6.10) and (6.11). This is what we call DiagonalizeBox in the algorithm of Fig. 7.1.[1]

7.3 Example

To see how the algorithm acts on a concrete example, we refer to the graph on the left of Fig. 7.4. A parse tree for a slick 2-expression representing this graph is found at the right of Fig. 7.4, where we label the internal nodes as A, B, C, D, E, and F. In the expressions, *id* stands for the identity function, and $\{1 \rightarrow 2\}$ denotes the

---

[1] As it turns out, DiagonalizeBox is a slight variation of the algorithm in [4, 5], but the proof follows the same arguments.

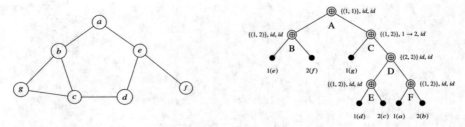

Fig. 7.4 A graph with $k = 2$ and its parse tree

function that maps 2 to 1 and keeps the other elements unchanged. Since G is not a cograph (it contains an induced copy of P_4), we know that $scw(G) = 2$.

We apply the algorithm to the graph defined by this slick 2-expression for $x = 0$. Since $k = 2$ and $x = 0$, the boxes created by the leaves may be of the following two types:

$$1(v) : [k', k'', M, labels] = [0, 1, (0), (1)], \quad \text{or } 2(v) : [0, 1, (0), (2)].$$

This means that $k' = 0$ (thus $M^{(0)}$ is empty) and that $k'' = 1$ and $M^{(2)} = (0)$.

The nodes B, E, and F of the parse tree perform identical operations to produce

$$\left[0, 2, \begin{pmatrix} 0 & 1 \\ 1 & 0 \end{pmatrix}, \begin{pmatrix} 1 \\ 2 \end{pmatrix} \right], \tag{7.8}$$

meaning that $M^{(0)}$ is empty ($k' = 0$), $k'' = 2$, and $M^{(2)} = \begin{pmatrix} 0 & 1 \\ 1 & 0 \end{pmatrix}$, and the labels of its rows are 1 and 2, respectively. Up to this point, we did not need to apply the congruence operations described above to insure type II rows have unique labels or to insert type I rows into $M^{(0)}$.

Next, we process node D that receives boxes from nodes E and F given by Eq. (7.8) that, together with $S = \{(2, 2)\}$, $L = id$, and $R = id$ produce the following initial matrix and vector of labels:

$$\begin{pmatrix} 0 & 1 & 0 & 0 \\ 1 & 0 & 0 & 1 \\ 0 & 0 & 0 & 1 \\ 0 & 1 & 1 & 0 \end{pmatrix}, \quad \begin{pmatrix} 1 \\ 2 \\ 1 \\ 2 \end{pmatrix}.$$

We notice that rows 1 and 3 are type II with the same label. We then perform the operations as in Eq. (7.6), given by $R_1 \leftarrow R_1 - R_3$, $C_1 \leftarrow C_1 - C_3$, leading to the matrix

$$
\begin{pmatrix}
0 & 1 & 0 & -1 \\
1 & 0 & 0 & 1 \\
0 & 0 & 0 & 1 \\
-1 & 1 & 1 & 0
\end{pmatrix},
\begin{pmatrix}
1 \\
2 \\
1 \\
2
\end{pmatrix}.
$$

The dotted lines separate the type I rows and columns from the type II rows and columns.

We need to insert the type i row into the empty matrix $M^{(0)}$. Since x_v is empty, $d_v = 0 = M^{(0)}$, we have the matrix $M^{(1)} = y_v = [1, 0, -1]$ that is already in row echelon form.

Now, as rows 2 and 4 have equal labels, we perform $R_2 \leftarrow R_2 - R_4$, $C_2 \leftarrow C_2 - C_4$, as in Eq. (7.6), so that the type ii rows have unique labels, arriving at the matrix

$$
\begin{pmatrix}
0 & 2 & 0 & -1 \\
2 & -2 & -1 & 1 \\
0 & -1 & 0 & 1 \\
-1 & 1 & 1 & 0
\end{pmatrix},
\begin{pmatrix}
1 \\
2 \\
1 \\
2
\end{pmatrix}.
$$

In order to insert the new type I row into matrix $[M^{(0)}, M^{(1)}] = [0, 2, 0, -1]$, we exchange rows/columns 1 and 2, obtaining

$$
\begin{pmatrix}
-2 & 2 & -1 & 1 \\
2 & 0 & 0 & -1 \\
-1 & 0 & 0 & 1 \\
1 & -1 & 1 & 0
\end{pmatrix},
\begin{pmatrix}
2 \\
1 \\
1 \\
2
\end{pmatrix}.
$$

The new row to be inserted is now the first row, with $d_v = -2$, $x_v = [2]$, and $y_v = [-1, 1]$, meaning that we are in the subcase 2 of procedure CombineBoxes. As x_v has a single nonzero entry, there is no need to eliminate any other entry, and since $d_v = -2 \neq 0$, we execute the operations in (6.10) with $\alpha_j = 2$, namely, $R_1 \leftarrow R_1 + \frac{1}{2}R_2$, and $C_1 \leftarrow C_1 + \frac{1}{2}C_2$, obtaining

$$
\begin{pmatrix}
0 & 2 & -1 & 1/2 \\
2 & 0 & 0 & -1 \\
-1 & 0 & 0 & 1 \\
1/2 & -1 & 1 & 0
\end{pmatrix},
\begin{pmatrix}
2 \\
1 \\
1 \\
2
\end{pmatrix}.
$$

Execute now the operations of Eq. (6.11) with rows 1 and 2, that is $R_2 \leftarrow R_2 + \frac{1}{2}R_1$, $C_2 \leftarrow C_2 + \frac{1}{2}C_1$, followed by $R_1 \leftarrow R_1 - R_2$, and $C_1 \leftarrow C_1 - C_2$, giving

$$\begin{pmatrix} -2 & 0 & -1/2 & 5/4 \\ 0 & 2 & -1/2 & -3/4 \\ -1/2 & -1/2 & 0 & 1 \\ 5/4 & -3/4 & 1 & 0 \end{pmatrix}, \begin{pmatrix} 1 \\ 2 \\ 1 \\ 2 \end{pmatrix}.$$

As in (6.7), we use -2 to diagonalize its row/column and 2 to diagonalize its row/column, obtaining two new diagonal values. We then add the values 2 and -2 to the diagonal D. The box returned by node D is

$$\left[0, 2, \begin{pmatrix} 0 & 1/2 \\ 1/2 & 1/2 \end{pmatrix}, \begin{pmatrix} 1 \\ 2 \end{pmatrix} \right].$$

When the algorithm starts processing node C, it uses the boxes produced by node D and a leaf with label 1, together with $S = \{(1, 2)\}$. The vertex of the left component is relabeled $1 \rightarrow 2$. This leads to

$$\begin{pmatrix} 0 & 0 & 1 \\ 0 & 0 & 1/2 \\ 1 & 1/2 & 1/2 \end{pmatrix}, \begin{pmatrix} 2 \\ 1 \\ 2 \end{pmatrix}.$$

At this point, $M^{(2)}$ is a 3×3 matrix and $k'' = 3 > k = 2$. We first ensure the uniqueness of labels for type II. The operations are $R_1 \leftarrow R_1 - R_3$ $C_1 \leftarrow C_1 - C_3$, leading to

$$M = \begin{pmatrix} -3/2 & -1/2 & 1/2 \\ -1/2 & 0 & 1/2 \\ 1/2 & 1/2 & 1/2 \end{pmatrix} \text{ with labels } \begin{pmatrix} 2 \\ 1 \\ 2 \end{pmatrix}.$$

Now we need to insert the first type I row into the empty matrix $M^{(0)}$. As x_v is empty, $d_v = -3/2$, and it is subcase 1c, we can use it to diagonalize this row/column, arriving at

$$M = \begin{pmatrix} -3/2 & 0 & 0 \\ 0 & 1/6 & 1/3 \\ 0 & 1/3 & 2/3 \end{pmatrix}.$$

We obtain a new diagonal value $-3/2$, and the box returned by node C is

$$\left[0, 2, \begin{pmatrix} 1/6 & 1/3 \\ 1/3 & 2/3 \end{pmatrix}, \begin{pmatrix} 1 \\ 2 \end{pmatrix} \right].$$

We finally process node A, which combines the boxes produced by B, which is given by Eq. (7.8), and C. Note that, in this matrix, we have $k' = 0$, $k'' = 4$, and after adding the edges of $S = \{(1, 1)\}$, we have

$$\begin{pmatrix} 0 & 1 & 1 & 0 \\ 1 & 0 & 0 & 0 \\ 1 & 0 & 1/6 & 1/3 \\ 0 & 0 & 1/3 & 2/3 \end{pmatrix} \text{ with labels } \begin{pmatrix} 1 \\ 2 \\ 1 \\ 2 \end{pmatrix}.$$

Applying the operations to guarantee the uniqueness of labels, namely $R_1 \leftarrow R_1 - R_3$ and $C_1 \leftarrow C_1 - C_3$, we obtain the following matrix:

$$\begin{pmatrix} -11/6 & 1 & 5/6 & -1/3 \\ 1 & 0 & 0 & 0 \\ 5/6 & 0 & 1/6 & 1/3 \\ -1/3 & 0 & 1/3 & 2/3 \end{pmatrix} \text{ with labels } \begin{pmatrix} 1 \\ 2 \\ 1 \\ 2 \end{pmatrix}.$$

To insert the first row into the empty matrix $M^{(0))}$, we observe that $d_v = -11/6 \neq 0$, which is subcase 1c of procedure CombineBoxes. We use d_v to diagonalize this row, obtaining the diagonal value $-11/6$ and the new matrix

$$M = \frac{1}{11} \begin{pmatrix} 6 & 5 & -2 \\ 5 & 6 & 2 \\ -2 & 2 & 8 \end{pmatrix}, \text{ with labels } \begin{pmatrix} 2 \\ 1 \\ 2 \end{pmatrix}.$$

Since rows 1 and 3 have the same label, we execute $R_1 \leftarrow R_1 - R_3$ $C_1 \leftarrow C_1 - C_3$, which produce the matrix

$$M = \frac{1}{11} \begin{pmatrix} 18 & 3 & -10 \\ 3 & 6 & 2 \\ -10 & 2 & 8 \end{pmatrix}, \text{ with labels } \begin{pmatrix} 2 \\ 1 \\ 2 \end{pmatrix}.$$

We use the value $d_v = 18/11$ to diagonalize this row and column, obtaining the diagonal value $18/11$. The remaining matrix is

$$M = \begin{pmatrix} 1/2 & 1/3 \\ 1/3 & 2/9 \end{pmatrix},$$

giving the diagonal values 0 and $1/2$.

The final diagonal values are $D = (-2, 2, -\frac{3}{2}, -\frac{11}{6}, \frac{18}{11}, \frac{1}{2}, 0)$. This means that the graph has three negative eigenvalues, three positive eigenvalues, and 0 is an eigenvalue with multiplicity one. In fact, the actual spectrum may be approximated by the multiset

$$\{-1.93, -1.61, -1, 0, 0.62, 1.46, 2.47\}.$$

7.4   Correctness, Complexity, and Implementation

In this section, we use the results from the previous section to prove the correctness of the algorithm. Moreover, we give the complexity of the algorithm by studying the number of operations performed.

Theorem 7.1 *Let G be a graph with adjacency matrix A, given by a slick expression Q_G with parse tree T, and let $c \in \mathbb{R}$. Algorithm* Diagonalize Clique-width *correctly outputs the diagonal elements of a diagonal matrix congruent to $B = A + xI$. Moreover, this is done in $O(k^2 n)$ operations.*

Proof It is clear that the operations performed by Diagonalize Clique-width are congruence operations. Indeed, besides simultaneous permutations of rows and columns, each elementary row operation performed is followed by the same elementary column operation. We now argue that the elements that the algorithm outputs are precisely the diagonal elements of a diagonal matrix congruent to $B = A + xI$. As above, let $\widetilde{B}_{Q_t}^+$ be the $n \times n$ matrix obtained from B by performing the same congruence operations on B that are actually performed on the corresponding rows of matrices M_{Q_t} up to stage t. Lemma 7.2 ensures that, at the beginning of each stage t, the algorithm always starts from a submatrix M of $\widetilde{B}_{Q_{t-1}}^+$. Also, Lemma 7.1 guarantees that Eq. (7.3) holds at the end of each stage: part (a) ensures that a row that was diagonalized at some stage cannot be modified at later stages. If $v \in G(Q_t)$ and $w \notin G(Q_t)$ (and has not been diagonalized), part (b) ensures (applying it twice with the roles of v and w interchanged) that the element β_i^j corresponding to the entry vw in $\widetilde{B}_{Q_t}^+$ is equal to 0 if v has type I or if w has type I in M_{Q_τ} for some $\tau < t$. Finally, part (c) ensures that in the remaining cases the entry vw is equal to the corresponding entry in B. We conclude that, at each stage, the elements in the output of CombineBoxes (which are called *diagonals* in Fig. 7.3) are indeed diagonal elements of the final matrix produced. The only nonzero elements of the final matrix that are not of this form are obtained by the application of DiagonalizeBox, which clearly outputs the final diagonal elements of a diagonal matrix D congruent to B.

The bound of $O(k^2 n)$ is obtained by a clever accounting for the time spent by executing the elementary operations. The elementary operations to maintain rows in $M^{(2)}$ with a single label can be done in $O(k)$. The matrix $M^{(1)}$ is maintained in row echelon form, and inserting one row vector incurs a cost of $O(k^2)$. It does not matter that up to k row vectors are inserted in one node because every row vector is inserted only once, and there are only n rows, one for each vertex.   □

By applying Corollary 3.1, in two calls of `Diagonalize Clique-width`, we can determine the number of eigenvalues of a graph G with clique-width k in any interval.

Corollary 7.1 *The number of eigenvalues of A in a real interval I can be computed in time $O(k^2 n)$ for graphs of clique-width k.*

Finally, we discuss some implementation issues of `Diagonalize Clique-width`, by itemizing a few features of our algorithm, that we added to simplify the description but are not necessary in an efficient implementation. It is worth noticing that the whole algorithm is very fast, as there are no large constants hidden in the O notation.

(a) Since the matrix $\widetilde{B_Q^+}$ is not computed, one may easily write `Diagonalize Clique-width` as a recursive algorithm.
(b) It is not necessary to perform permutations of rows and columns to separate them according to type, and it suffices to keep track of the vertices of each type in matrices M_{Q_t}.
(c) In the final step at the root, we could map all the vertices to the same label, essentially producing an unlabeled graph.

As was the case for the previous chapter, we end this chapter by stating our belief that the algorithm presented here will be useful as a tool to derive results about spectral parameters, particularly when the algorithm is applied to graph classes with bounded clique-width and whose clique decompositions are well-understood. A good illustration is Chap. 8, where we describe an algorithm for a superclass of cographs, whose elements are called distance-hereditary graphs and may be described by algebraic expressions known as pruning sequences. Other classes for which this approach may be fruitful are classes whose elements may be characterized by a well-structured parse tree. For example, this is the case for P_4-reducible [8], P_4-sparse [7], and matrogenic [3] graphs. For a general survey of graph classes, we refer to [1]. Parse trees for graphs in a graph class may often be derived from their modular decompositions, see [6]. A compilation of results about the clique-width of hereditary graph classes may be found in [2].

References

1. Brandstädt, A., Le, V.B., Spinrad, J.P.: Graph Classes: A Survey. Society for Industrial and Applied Mathematics (1999). https://doi.org/10.1137/1.9780898719796
2. Dabrowski, K.K., Johnson, M., Paulusma, D.: Clique-width for hereditary graph classes, pp. 1–56. London Mathematical Society Lecture Note Series. Cambridge University Press (2019)
3. Földes, S., Hammer, P.L.: On a class of matroid-producing graphs. Combinatorics, Keszthely 1976, Colloq. Math. Soc. Janos Bolyai **18**, 331–352 (1978)
4. Fürer, M., Hoppen, C., Jacobs, D.P., Trevisan, V.: Locating the eigenvalues for graphs of small clique-width. In: M.A. Bender, M. Farach-Colton, M.A. Mosteiro (eds.) LATIN 2018: Theoretical Informatics, pp. 475–489. Springer International Publishing, Cham (2018)

5. Fürer, M., Hoppen, C., Jacobs, D.P., Trevisan, V.: Eigenvalue location in graphs of small clique-width. Linear Algebra Appl. **560**, 56–85 (2019)
6. Habib, M., Paul, C.: A survey of the algorithmic aspects of modular decomposition. Comput. Sci. Rev. **4**(1), 41–59 (2010)
7. Hoàng, C.: A class of perfect graphs. Master's thesis, McGill University (1983)
8. Jamison, B., Olariu, S.: P_4-reducible graphs - a class of uniquely tree-representable graphs. Stud. Appl. Math. **81**(1), 79–87 (1989)

Chapter 8
Distance-Hereditary Graphs

8.1 Distance-Hereditary Graphs

A graph G is called *distance-hereditary* if any induced path is isometric, that is, if $d_G(x, y) = d_H(x, y)$ holds for every connected induced subgraph H of G and for all vertices $x, y \in V(H)$. Distance-hereditary graphs were introduced by Howorka [6] and have also been called *completely separable* in [5].

Linear time diagonalization of distance-hereditary graphs comes from the linear time construction of pruning trees (described below), a linear time translation of pruning trees into a slick 2-expression, and the application of the linear time algorithm in Chap. 7.

It is easy to see that the graphs in Fig. 8.1 are not distance-hereditary. For example in the leftmost graph, removing the bottom vertex changes the distance between the leftmost and rightmost vertices. Similar arguments apply to the other two graphs. We will see in the next section that the graph in Fig. 8.2 is distance-hereditary.

Clearly, any tree is distance-hereditary. Indeed, any two vertices in a tree are connected by a single path, and their distance is given by the length of this path. Therefore, if x and y lie in a subtree H of a tree T, then the single path between x and y in T also belongs to H, and their distance is the same. In the next section, we show that all cographs are distance-hereditary.

The class of distance-hereditary graphs can be characterized in many ways. We list only some of these characterizations, and others can be found in the papers referenced. In [1], Bandelt and Mulder show that in a connected graph each of the following conditions is equivalent to G being distance-hereditary. Condition (c) implies that distance-hereditary graphs are hereditary in the sense of Sect. 4.1.

(a) For any vertices u and v in G with $d_G(u, v) = 2$, there is no induced path from u to v having length greater than two.

© The Author(s), under exclusive license to Springer Nature Switzerland AG 2022
C. Hoppen et al., *Locating Eigenvalues in Graphs*, SpringerBriefs in Mathematics,
https://doi.org/10.1007/978-3-031-11698-8_8

Fig. 8.1 Graphs that are not distance-hereditary

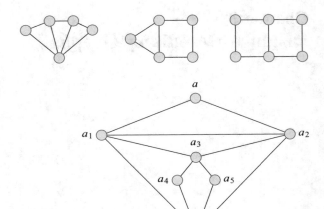

Fig. 8.2 The graph constructed with the pruning sequence in (8.1)

(b) For any four vertices u, v, w, x in G, at least two of the sums are equal

$$d_G(u, v) + d_G(w, x), \quad d_G(u, w) + d_G(v, x), \quad d_G(u, x) + d_G(v, w).$$

(c) G is defined by Forb($C \cup \mathcal{D}$), where $\mathcal{D}$ contains the three graphs in Fig. 8.1 and C consists of cycles C_n, $n \geq 5$. That is, they are $\mathcal{F}$-free , where $\mathcal{F} = C \cup \mathcal{D}$.

Another approach is a construction presented in [1, 4, 5] for distance-hereditary graphs and is defined as follows. Given a graph G with vertex set $V = \{v_1, \ldots, v_n\}$, let $S = \{s_2, \ldots, s_n\}$ be a sequence of pairs, each s_i having one of the following three forms:[1]

$$\langle (v_i, v_j), \boldsymbol{\ell} \rangle, \quad \langle (v_i, v_j), \boldsymbol{f} \rangle \quad \text{or} \quad \langle (v_i, v_j), \boldsymbol{t} \rangle,$$

where $i > j$. We say that S is a *pruning sequence* for G if, for all i, $2 \leq i \leq n$, whenever $s_i = \langle (v_i, v_j), \boldsymbol{\ell} \rangle$, $s_i = \langle (v_i, v_j), \boldsymbol{f} \rangle$ or $s_i = \langle (v_i, v_j), \boldsymbol{t} \rangle$, the subgraph of G induced by $\{v_1, \ldots, v_i\}$ is obtained from the subgraph induced by $\{v_1, \ldots, v_{i-1}\}$, by making the vertex v_i a leaf, a duplicate, or a coduplicate, respectively, of v_j. Consider the pruning sequence for a graph with vertex set $\{a, a_1, a_2, a_3, a_4, a_5, a_6\}$:

$$\langle (a_1, a), \boldsymbol{\ell} \rangle, \ \langle (a_2, a), \boldsymbol{t} \rangle, \ \langle (a_3, a), \boldsymbol{f} \rangle, \ \langle (a_4, a_3), \boldsymbol{\ell} \rangle, \ \langle (a_5, a_3), \boldsymbol{\ell} \rangle, \ \langle (a_6, a_3), \boldsymbol{f} \rangle.$$
$$(8.1)$$

---

[1] We adopt the notation in [4, 5] where $\boldsymbol{f}$ stands for *false* twin and $\boldsymbol{t}$ represents *true* twin. Recall from Chap. 4 that false twins and true twins have the same meanings, respectively, as duplicates and coduplicates.

This pruning sequence constructs the graph shown in Fig. 8.2. Bandelt and Mulder in [1] and Hammer and Maffray in [5] independently showed the following characterization.

Theorem 8.1 *A connected graph G is distance-hereditary if and only if G has a pruning sequence.*

8.2   Locating Eigenvalues in Distance-Hereditary Graphs

The purpose of this section is to prove that distance-hereditary graphs have a linear time diagonalization algorithm like the other graph classes in this book. Unlike other chapters, we will not exhibit pseudocode for the algorithm but rather show that these graphs have a linear time diagonalization algorithm with the structure of the algorithm in Chap. 7. In doing so, we also show that these graphs have slick clique-width at most 2, the parameter described in Sect. 4.6. A nice feature of our proof is that a slick 2-expression may be derived from a pruning sequence of G. In case the pruning sequence has no terms with an ℓ, then the graph is a cograph, and a slick 1-expression is produced.

Theorem 8.2 *For every distance-hereditary graph G, $scw(G) \leq 2$.*

Before proving this, we show that trees have this property. Since trees are distance-hereditary, this will follow from Theorem 8.2. However, the proof is sufficiently short and elegant to justify it, and it is related to the general result.

Proposition 8.1 *If T is a tree, $scw(T) \leq 2$.*

Proof By induction on n, we show that any rooted tree can be constructed with a slick 2-expression in which the root r is labeled 1 and all other vertices are labeled 2. When $n = 1$, the expression is $1(r)$. Now assume $n > 1$ and that all trees of order $k < n$ have slick 2-expressions with the desired property. Let T be a tree of order $n > 1$ with root r. Then let $T_1, \ldots, T_j$ be the subtrees of T obtained by removing r. Root each T_i at r_i, the neighbor of r. By induction, each T_i has a slick 2-expression e_i that leaves r_i labeled 1 and all other vertices labeled 2. Let e' be the expression that forms the disjoint union of these trees, without altering their labels. The slick 2-expression $e' \oplus_{\{(1,1)\}, 1 \to 2, id} 1(r)$ produces the desired labeled tree.   □

The proof of Theorem 8.2 is modeled on a theorem by Golumbic and Rotics [4] in which they showed:

Theorem 8.3 *For every distance-hereditary graph G, $cw(G) \leq 3$.*

Note that, in combination with Theorem 4.6, Theorem 8.3 implies that $scw(G) \leq 3$ for every distance-hereditary graph G. We will show how modifying its proof improves the upper bound to 2.

The proof in [4] made use of a rooted plane tree representation of the pruning sequence, called the *pruning tree*, constructed as follows. Let $G = (V, E)$, where

Fig. 8.3 Pruning tree of
depth 2 that constructs the
graph in Fig. 8.2

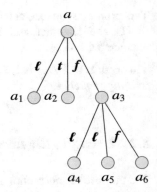

$V = \{v_1, \ldots, v_n\}$, and let $S = \{s_2, \ldots, s_n\}$ be a pruning sequence for G. Initially, $T = T_1$ is the tree consisting of a single vertex, root v_1. Given the tree T_{i-1}, if $s_i = \langle (v_i, v_j), \ell \rangle$ (respectively, $s_i = \langle (v_i, v_j), f \rangle$, or $s_i = \langle (v_i, v_j), t \rangle$), then the tree T_i is obtained by adding the vertex v_i as the rightmost child of v_j and labeling the edge with ℓ (respectively, f or t). The tree $T = T_n$ is the pruning tree of G. Figure 8.3 depicts a pruning tree with depth 2 formed from the pruning sequence in (8.1).

We say v is an *ℓ-child* of u in G if v is a child of u in T, and the edge between them is labeled ℓ. Similar definitions hold for *f-child* and *t-child*. We call v a *twin descendant* of u if, in T, either $v = u$, or v is a descendant of u, and all edges between them are labeled f or t. In Fig. 8.3, the nodes a, a_2, a_3, and a_6 are twin descendants of a.

If a is a node in T, then T_a denotes the subtree rooted at a. If $T_{a_1}, \ldots, T_{a_k}$ are disjoint subtrees of T, then $G[T_{a_1} \cup \cdots \cup T_{a_k}]$ denotes the subgraph of G induced by the vertices in $T_{a_1} \cup \cdots \cup T_{a_k}$. Recall that an *internal* node is a non-leaf. The following lemma, whose proof can be found in [4, Lemma 3.8], is critical to the proof of Theorem 8.3, as well as ours.

Lemma 8.1 *Let G be a graph having a pruning sequence S with corresponding pruning tree T, and let a be an internal node in T whose children, left to right, are $a_1, \ldots, a_\ell$. For $1 \le i \le \ell$, we have the following:*

(a) *If a_i is an f-child of a, then $G[\{a\} \cup T_{a_i} \cup T_{a_{i+1}} \cup \cdots \cup T_{a_\ell}]$ is the disjoint union of the graphs $G[T_{a_i}]$ and $G[\{a\} \cup T_{a_{i+1}} \cup \ldots \cup T_{a_\ell}]$.*

(b) *If a_i is either an ℓ-child or a t-child of a, then $G[\{a\} \cup T_{a_i} \cup T_{a_{i+1}} \cup \cdots \cup T_{a_\ell}]$ can be constructed by taking the disjoint union of $G[T_{a_i}]$ and $G[\{a\} \cup T_{a_{i+1}} \cup \cdots \cup T_{a_\ell}]$, and connecting all twin descendants of a_i to a, and to all twin descendants of a occurring in $T_{a_{i+1}} \cup \cdots \cup T_{a_\ell}$.*

Lemma 8.1 gives a formula for constructing the graph $G[T_a]$, right to left, if we know $G[T_{a_i}]$ for all $i \in \{1, \ldots, \ell\}$. It is understood that when $i = \ell$, the term $T_{a_{i+1}} \cup \cdots \cup T_{a_\ell}$ is empty. Note that when a_i is an f-child of a, adding the component is simple. However, when a_i is either an ℓ-child or a t-child, we must know where the twin descendants are.

In Fig. 8.3, by Lemma 8.1 part (a) , since a_3 is an f-child of a, $G[\{a\} \cup T_{a_3}] = G[\{a, a_3, a_4, a_5, a_6\}]$ is the disjoint union of $G[\{a\}]$ and $G[T_{a_3}] = G[\{a_3, a_4, a_5, a_6\}]$. By part (b), since a_1 is an ℓ-child of a, $G = G[\{a\} \cup T_{a_1} \cup T_{a_2} \cup T_{a_3}]$ can be constructed by taking the disjoint union of $G[\{a\} \cup T_{a_2} \cup T_{a_3}] = G[\{a, a_2, a_3, a_4, a_5, a_6\}]$ and $G[T_{a_1}] = G[\{a_1\}]$ and connecting a_1 to a, and connecting a_1 to a_2, a_3 and a_6. These facts can be verified in Fig. 8.2.

Proof of Theorem 8.2 We may assume that G is connected. By Theorem 8.1, G has a pruning sequence with pruning tree T. We show that for any node a in T, there is a slick 2-expression t_a that constructs the labeled graph $G[T_a]$ such that all twin descendants of a are labeled 1, and all other nodes are labeled 2. Our result follows choosing a as the root of T. We use induction on the depth d of T_a.

When $d = 0$, a is a leaf of T and the expression $1(a)$ suffices. To complete the induction, let T_a have depth $d > 0$, and let $a_1, a_2, \ldots, a_\ell$ be the children of a, left to right. By an induction assumption, there are slick 2-expressions $t_{a_1}, \ldots, t_{a_\ell}$ such that each t_{a_i} constructs $G[T_{a_i}]$ and the twin descendants of a_i are precisely the vertices labeled 1. We claim that

$$t_a = t_{a_1} \oplus_{S_1, L_1, R_1} \left(t_{a_2} \oplus_{S_2, L_2, R_2} (\cdots (t_{a_\ell} \oplus_{S_\ell, L_\ell, R_\ell} 1(a)) \cdots) \right) \tag{8.2}$$

is the required slick 2-expression, where

$$(S_i, L_i, R_i) = \begin{cases} (\emptyset, id, id) & \text{if } a_i \text{ is an } f\text{-child} \\ (\{(1, 1)\}, id, id) & \text{if } a_i \text{ is a } t\text{-child} \\ (\{(1, 1)\}, 1 \to 2, id) & \text{if } a_i \text{ is a } \ell\text{-child.} \end{cases} \tag{8.3}$$

Note that in (8.3), edges are placed between vertices having label 1, unless a_i is an f-child. Labels of the right operand never change. Labels of the left operand do not change unless a_i is an ℓ-child, in which case they become 2.

To prove the correctness of (8.2), we can argue right to left. Let us say that the 2-labeled graph $G[\{a\} \cup T_{a_i} \cup \cdots \cup T_{a_\ell}]$ is *correctly* labeled if the twin descendants of a are precisely the vertices with label 1. We first prove that $t_{a_\ell} \oplus_{S_\ell, L_\ell, R_\ell} 1(a)$ constructs and correctly labels $G[\{a\} \cup T_{a_\ell}]$. There are three cases:

If a_ℓ is an f-child, then $t_{a_\ell} \oplus_{\emptyset, id, id} 1(a)$ creates the disjoint union of the graph $G[T_{a_\ell}]$ and $G[\{a\}]$. By Lemma 8.1 part (a), this equals $G[T_{a_\ell} \cup \{a\}]$. Note that a is labeled 1. Since twin descendants of a_ℓ will remain twin descendants of a, the labels are correct.

If a_ℓ is a t-child, by Lemma 8.1 part (b), to obtain $G[T_{a_\ell} \cup \{a\}]$, we must add edges between a and all twin descendants of a_ℓ. This is accomplished by $t_{a_\ell} \oplus_{\{(1,1)\}, id, id} 1(a)$. The labels are correct as twin descendants of a_ℓ become twin descendants of a.

If a_ℓ is an ℓ-child, we have $t_{a_\ell} \oplus_{\{(1,1)\}, 1 \to 2, id} 1(a)$. By Lemma 8.1 part (b), we must also add edges as in the previous case. But since there is an ℓ on the edge above a_ℓ, no twin descendant of a_ℓ is a twin descendant of a, so labels in $G[T_{a_\ell}]$ are mapped to 2.

To complete the proof, now assume that

$$t' = t_{a_{i+1}} \oplus_{S_{i+1}, L_{i+1}, R_{i+1}} (\cdots (t_{a_\ell} \oplus_{S_\ell, L_\ell, R_\ell} 1(a)) \cdots)$$

generates and correctly labels the graph

$$G_{i+1} = G[\{a\} \cup T_{a_{i+1}} \cup \cdots \cup T_{a_\ell}].$$

It suffices to show that $t_{a_i} \oplus_{S_i, L_i, R_i} t'$ constructs and correctly labels

$$G_i = G[\{a\} \cup T_{a_i} \cup T_{a_{i+1}} \cup \cdots \cup T_{a_\ell}].$$

There are three cases depending on whether a_i is an f-child, t-child, or ℓ-child.

When a_i is an f-child, by Lemma 8.1 part (a), we need to only form a disjoint union. This is achieved with $t_{a_i} \oplus_{\emptyset, id, id} t'$. No labels are changed since all twin descendants of a_i remain twin descendants of a.

When a_i is a t-child, by Lemma 8.1 part (b), to obtain G_i, we must form $G[T_{a_i}] \cup G_{i+1}$ and then connect the twin descendants of a_i to a and to the twin descendants of a occurring in G_{i+1}. Note also that no twin descendants can change, and hence, labels should not be changed. This is achieved by $t_{a_i} \oplus_{\{(1,1)\}, id, id} t'$.

The case when a_i is an ℓ-child is the most subtle. By Lemma 8.1 part (b), we must form $G[T_{a_i}] \cup G_{i+1}$ and then connect the twin descendants of a_i to a and to the twin descendants of a occurring in G_{i+1}. However, since the edge above a_i is labeled ℓ, there can no longer be twin descendants of a in this subtree. This is obtained by $t_{a_i} \oplus_{\{(1,1)\}, 1 \to 2, id} t'$. It is crucial that relabeling occurs after the edges are placed, and this is a property of the slick operator $\oplus$. It follows that the slick 2-expression in (8.2) correctly labels and constructs the graph $G[T_a]$, completing the proof.   □

The expression t_a could have been defined recursively but we preferred to give an explicit formula. We give an example of the slick 2-expression constructed by Theorem 8.2. Consider the graph of order 4 having the pruning tree shown in Fig. 8.4. This is a cograph as it has no induced P_4. Then Formula (8.2) yields the slick 2-expression

$$1(a_1) \oplus_{\{(1,1)\}, 1 \to 2, id} \left(1(a_2) \oplus_{\{(1,1)\}, id, id} \left(1(a_3) \oplus_{\emptyset, id, id} 1(a) \right) \right).$$

Fig. 8.4 A small pruning tree that constructs a cograph

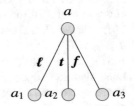

This constructs the graph in which a, a_2, a_3 are labeled 1 and a_1 is labeled 2. By Proposition 4.4, cographs have slick clique-width 1, and so the construction does not guarantee an expression with minimum labels if we use *any* pruning tree. The graph in Fig. 8.2 is not a cograph since it has an induced P_4, and so we cannot produce a slick 1-expression for this graph.

There are two interesting extremal cases in the proof of Theorem 8.2. Recall from Theorem 4.4 that G is a cograph if and only if it can be obtained from a sequence of sibling additions. This means G has a pruning tree having only t-children or f-children. In this case, formula (8.3) constructs a minimal slick 1-expression. This happens because the pruning tree has no ℓ-children, and therefore no relabeling occurs. Consequently, the label 2 is never used.

The other case occurs with trees that can be constructed with only ℓ-children. In this case, the slick 2-expression is *exactly* that described in Proposition 8.1.

All distance-hereditary graphs are perfect graphs [6]. We have shown that distance-hereditary graphs have bounded slick clique-width. This cannot be generalized to perfect graphs since permutation graphs and unit interval graphs, which are subclasses of perfect graphs, have unbounded clique-width [4] and must also have unbounded slick clique-width by Theorem 4.6.

We can now deduce the existence of a linear time eigenvalue location algorithm for the adjacency matrix of distance-hereditary graphs. Given a graph G, the authors in [3] showed that the pruning tree T of a distance-hereditary graph can be constructed in linear time. The proof of Theorem 8.2 shows that a pruning tree can be translated into a slick clique-width expression s for G in linear time. Given the expression s, we now can use the linear time diagonalization algorithm in Chap. 7.

8.3   The Graphs Having *scw* ≤ 2

In this section, we calculate the slick clique-width of all cycles and conclude that not all graphs having slick clique-width 2 are distance-hereditary. We also show that the class of graphs $\mathcal{H}_2$ having slick clique-width at most two cannot be defined with a finite set of forbidden subgraphs. In doing so, we give some interesting properties of slick clique-width. For $k \geq 1$, let $\mathcal{S}_k$ denote the class of graphs whose slick clique-width is k. The join operator $\oplus$ is defined in Sect. 4.2.

Proposition 8.2 *For any $k \geq 1$, $\mathcal{S}_k$ is closed under the following operations:*

(a) Complement
(b) Disjoint union $\cup$
(c) Join $\oplus$

Proof Let G be a graph with edge set E. Let $\overline{G}$ denote the complement of G. We claim that $scw(\overline{G}) = scw(G)$. Let T_G be a parse tree for G. We construct a parse tree $T_{\overline{G}}$ for $\overline{G}$ as follows. Replace each operator $\oplus_{S,L,R}$ in T_G with $\oplus_{\overline{S},L,R}$, where $\overline{S}$ is the set of label pairs not in S. Since, for each vertex pair $\{u, v\}$, there is exactly one node in the parse tree where the edge $\{u, v\}$ may be added, namely the lowest common ancestor of the leaves corresponding to u and v, this places edges precisely on $\overline{E}$. Therefore, $scw(\overline{G}) \leq scw(G)$. Repeating the argument shows

$$scw(G) = scw(\overline{\overline{G}}) \leq scw(\overline{G}),$$

establishing (a).

Now let $G, H \in S_k$, having parse trees T_G and T_H of width k. We form a parse tree for $G \cup H$ by $T_G \oplus_{\emptyset, id, id} T_H$. This shows $scw(G \cup H) \leq k$. Since G is an induced subgraph of $G \cup H$, by Proposition 4.4(a), $k = scw(G) \leq scw(G \cup H) \leq k$, showing $G \cup H \in S_k$, establishing (b). If G and H are graphs, then $G \oplus H = \overline{\overline{G} \cup \overline{H}}$, so (c) follows from parts (a) and (b). □

The properties in Proposition 8.2 are not always shared with ordinary clique-width. In fact, a theorem of Courcelle and Olariu [2, Thr. 4.1] says that the clique-width of a graph's complement is bounded only by a factor of two. Here is a simple proof of their inequality using properties of slick clique-width.

Proposition 8.3 *If $cw(G) = k$, then $cw(\overline{G}) \leq 2k$.*

Proof Using Theorem 4.6, Proposition 8.2(a), and then Theorem 4.6, we have

$$cw(\overline{G}) \leq 2scw(\overline{G}) = 2scw(G) \leq 2cw(G) = 2k.$$

□

A similar argument using Theorem 4.6 and Proposition 8.2(c) shows:

Proposition 8.4 *If $cw(G) = cw(H) = k$, then $cw(G \oplus H) \leq 2k$.*

For $k \geq 1$, let $\mathcal{H}_k$ denote the class of graphs G with $scw(G) \leq k$, that is, $\mathcal{H}_k = \cup_1^k S_i$. Proposition 4.4(a) shows that the classes $\mathcal{H}_k$ are hereditary. By Theorem 4.1, there is a family $\mathcal{F}_k$ such that $\mathcal{H}_k = \text{Forb}(\mathcal{F}_k)$. We can assume $\mathcal{F}_k$ are minimal members. Said another way, given k, $\mathcal{F}_k$ is the set of all graphs G such that $scw(G) > k$, but $scw(H) \leq k$ for any proper induced subgraph H of G.

Proposition 4.4(b) establishes that $\mathcal{H}_1$ is the class of cographs, and Theorem 4.2 tells us that $\mathcal{F}_1 = \{P_4\}$. A natural question would be to characterize $\mathcal{H}_k$ and $\mathcal{F}_k$ for other values of k.

It would be nice if $G \in \mathcal{H}_2$ implied that G is distance-hereditary. However, we shall see that this is not the case. We will show that the cycles C_5 and C_6 belong to $\mathcal{H}_2$, but as mentioned earlier they are not distance-hereditary. Moreover, we shall prove that $\mathcal{F}_2$ is infinite.

Lemma 8.2 *$G \in \mathcal{F}_k$ if and only if $\overline{G} \in \mathcal{F}_k$.*

Proof Suppose $G \in \mathcal{F}_k$. To show that $\overline{G} \in \mathcal{F}_k$, we must show $scw(\overline{G}) > k$ and $scw(H) \leq k$ for any induced proper subgraph H of $\overline{G}$. We know $scw(G) > k$. By Proposition 8.2(a), $scw(\overline{G}) = scw(G)$, so $scw(\overline{G}) > k$. Now let H be a proper induced subgraph of $\overline{G}$ on vertex set S. Let K be the proper induced subgraph of G on S. Then $scw(K) \leq k$. However, $H = \overline{K}$, and by Proposition 8.2(a), $scw(H) \leq k$.   □

It is interesting to observe that in the case of $\mathcal{F}_1$, the graphs P_4 and $\overline{P_4}$ are isomorphic.

The remainder of this chapter is devoted to proving Theorem 8.4 that determines the slick clique-width of all cycles. Often for brevity we will equate the slick expression Q with the labeled graph constructed by Q.

Lemma 8.3 $scw(C_3) = 1$ and $scw(C_4) = 1$.

Proof The cycles C_3 and C_4 are cographs.   □

Lemma 8.4 $scw(C_5) = 2$ and $scw(C_6) = 2$.

Proof The graph C_5 is not a cograph, yet can be constructed as $1(v) \oplus_{\{(1,1)\}, id, id}$ Q, where Q generates a P_4 whose ends are labeled 1, and whose internal vertices are labeled 2. Similarly, C_6 is not a cograph yet can be constructed as $Q_\ell \oplus_{\{(1,1),(2,2)\}, id, id} Q_r$, where Q_ℓ generates $2K_2$ such that each component has both labels, while Q_r generates two isolated vertices, one with each label.   □

Lemma 8.5 For $n \geq 7$, $scw(C_n) \leq 3$.

Proof Let $n \geq 7$. We claim that the cycle C_n can be created by a slick 3-expression. Indeed, start with a slick 3-expression Q for the path P_3, whose vertices are labeled 1,2,3 along the path. We may increase the path by performing the operation

$$Q \oplus_{\{(3,3)\}, 3 \rightarrow 2, id} 3(v),$$

so that we have a P_4 whose internal vertices are labeled 2 and whose ends are labeled 1 and 3. Repeating this operation, we may obtain any path of finite length whose internal vertices are labeled 2 and whose ends have the other two labels. It is then easy to close the cycle using a new vertex that is adjacent to both ends of the path.   □

Before completing the proof of Theorem 8.4, we first prove an auxiliary result. We use $V(Q)$ to denote the vertex set of a slick expression Q, and we use $\star$ to represent either value in the label set.

Lemma 8.6 For $m \geq 6$, there is no slick 2-expression that creates a path P_m, such that the internal vertices and endpoints have different labels.

Proof By contradiction, suppose that such a slick 2-expression Q exists, where

$$Q = Q_\ell \oplus_{S,L,R} Q_r$$

for nonempty Q_ℓ and Q_r. Since the graph is connected, at least one edge must be added by $\oplus_{S,L,R}$. Without loss of generality, we may assume $|V(Q_\ell)| \geq |V(Q_r)|$. Let us also assume that labels 1 and 2 occur t_1 and t_2 times in Q_ℓ, respectively, where $t_2 \leq t_1$. Then, since $m \geq 6$, either $t_1 \geq 3$ or $t_1 = 2$ and $t_2 \geq 1$.

We first consider the case $t_1 \geq 3$. Then no edge can appear because of some element $(1, \star) \in S$; otherwise, $\oplus$ would produce a vertex (originally in Q_r) of degree at least 3. Since their degree cannot change, the labeling of the final path implies that all vertices with label 1 must have degree 2 in the graph induced by Q_ℓ. Since this graph does not contain cycles, it contains at least one path P whose internal vertices have label 1 and whose endpoints have label 2. This leads to a contradiction, as some edge must be created by some element $(2, \star) \in S$, creating a cycle through P.

As a consequence, we must have $t_1 = 2$. Call u and v the vertices of Q_ℓ with label 1. There are two subcases. First suppose that some element of S produces edges incident with u and v, say $(1, 1) \in S$. This implies that there is a single vertex w of Q_r with label 1 (to avoid C_4). Also, w is an isolated vertex in the graph induced by Q_r (because of its degree in the final graph). We claim that *neither* the elements labeled 1 in Q_ℓ can be an endpoint in P_m. They cannot *both* be endpoints, as $m \geq 6$ and they have a common neighbor. The claim follows because u and v must have the same label in the final graph, and we are assuming that the labels of the endpoints and of the internal vertices differ. Hence, both u and v must have degree two in the final path. Since no edges can be produced by the presence of $(1, 2)$ in S (as this would lead to a cycle through u and v) and the final graph is connected, it must be that both u and v have a neighbor of label 2 in the graph defined by Q_ℓ, call them u' and v'. Note that $u' \neq v'$ otherwise $\oplus$ would produce a cycle. Our assumptions so far imply that $|V(Q_\ell)|$ is exactly four and that because $m \geq 6$, there is at least one vertex of label 2 in the graph produced by Q_r. By our construction up to this point, the presence of any element $(\star, 2)$ in S would lead to a cycle in the final graph. As no vertices of label 2 in Q_r can be connected to w in the graph produced by Q_r, the final graph is disconnected, a contradiction.

This leads us to the final subcase, where $\oplus$ does not produce edges incident with u and v; thus, we assume that $\{(1, 1), (1, 2)\} \cap S = \emptyset$. Let w be a vertex of label 2 in Q_ℓ, and without loss of generality, assume that $(2, 2) \in S$ and that there is a vertex w' of label 2 in Q_r. Note that u and v must lie in components of Q_ℓ that include a vertex of label 2; otherwise, Q would produce a disconnected graph. Since $|V(Q_\ell)| \leq 4$ in this case, there must be another vertex in Q_r (because $m \geq 6$). Moreover, any such vertex must have label 1 and $(2, 1) \notin S$ (otherwise, some vertex of label 2 in Q_ℓ would have degree at least three in the final graph). We claim that Q_r is connected. Indeed, as Q_r has a single 2-vertex, if Q_r were disconnected, it would have a component whose vertices are all labeled 1, and S would not connect it to the remainder of the graph. From this, we deduce that Q_ℓ has a single vertex of label 2; otherwise, w' would have degree at least three in the final graph, and so $|V(Q_\ell)| = |V(Q_r)| = 3$. To obtain P_6, both Q_ℓ and Q_r must each produce a P_3 with one of the endpoints labeled 2. However, in this situation, no functions L and R can produce the desired labeling. □

Theorem 8.4 *For cycles C_n, we have*

$$scw(C_n) = \begin{cases} 1, & \text{if } n \in \{3, 4\} \\ 2, & \text{if } n \in \{5, 6\} \\ 3 & \text{if } n \geq 7. \end{cases}$$

Proof Lemmas 8.3 and 8.4 handle the small cases, so we may assume $n \geq 7$. By Lemma 8.5, it suffices to show that there is no slick 2-expression for C_n if $n \geq 7$. Suppose by contradiction that such an expression exists, and write it as $Q = Q_\ell \oplus_{S,L,R} Q_r$ for nonempty Q_ℓ and Q_r. Since C_n is connected, at least one edge must be added by $\oplus_{S,L,R}$. Without loss of generality, assume that the label that appears most often in Q_ℓ is 1. Also assume $|V(Q_\ell)| \geq |V(Q_r)|$.

There are two cases. If label 1 appears three or more times, then no edges can be added between vertices of label 1 in Q_ℓ and vertices in Q_r; otherwise, a vertex of Q_r would end up with degree at least three. So all vertices with label 1 have degree 2 in Q_ℓ, and because some edge must be added by $\oplus_{S,L,R}$, there must be vertices of Q_ℓ with label 2. (Note that all vertices with label 2 must have the same degree in Q_ℓ.) In fact, since the maximum degree of the graph induced by Q_ℓ is 2, the components of this graph must be paths whose endpoints are labeled 2 and whose internal vertices are labeled 1. This now implies that there is only one such path; otherwise, some vertex of Q_r would have degree at least 4 after $\oplus_{S,L,R}$ is performed. The only possibility would be that Q_ℓ generates P_{n-1}, but Lemma 8.6 tells us that this cannot happen.

Now assume that label 1 appears exactly twice in Q_ℓ; hence, label 2 must also appear exactly twice because $n \geq 7$. Since $\oplus_{S,L,R}$ creates at least one edge, we may assume without loss of generality that $(1, 1) \in S$ and that there is a vertex with label 1 in Q_r. This implies that label 1 appears exactly once in Q_r for otherwise $\oplus$ would create a C_4. Thus label 2 appears two or three times because $|V(Q_\ell)| = 4$, $n \geq 7$, and $|V(Q_\ell)| \geq |V(Q_r)|$. Moreover, there can be no other element in S. Indeed, $(1, 2) \in S$ produces vertices of degree at least 3, and $(2, 1) \in S$ creates a vertex of degree 4. Finally, if $(2, 2) \in S$, the resulting graph has a C_4 if Q_r has two vertices with label 2, and if Q_r has three vertices with label 2 the graph has vertices of degree 3. So Q_r must induce a graph where vertices of label 2 have degree 2 and the vertex of label 1 has degree 0. This is impossible if $|V(Q_r)| = 3$. If $|V(Q_r)| = 4$, the graph contains a C_3, a contradiction. $\qquad \square$

Corollary 8.1 $\mathcal{F}_2$ *is infinite and contains C_n and $\overline{C_n}$ for all $n \geq 7$.*

Proof This follows because proper induced subgraphs of cycles are forests whose slick clique-width is at most two, and from Lemma 8.2. $\qquad \square$

There are other elements in $\mathcal{F}_2$ besides those listed in Corollary 8.1. It is not possible to produce C_5 with two edges attached to *consecutive* pendant vertices using a slick 2-expression. But we have no simple proof of this.

Open problems include characterizing the graphs having slick clique-width 2, that is, S_2. Note these are precisely the non-cographs in $\mathcal{H}_2$. One could also try characterizing the forbidden graphs $\mathcal{F}_2$. These are graphs $G \notin \mathcal{H}_2$ but $K \in \mathcal{H}_2$ for all proper induced subgraphs K of G. While clique-width and slick clique-width are linearly related, we have seen in Propositions 4.3 and 8.2 that slick expressions have certain elegant properties not shared with ordinary expressions. What other applications for slick expressions are there?

References

1. Bandelt, H.J., Mulder, H.M.: Distance hereditary graphs. J. Comb. Theory Ser. B **41**(2), 182–208 (1986). https://doi.org/10.1016/0095-8956(86)90043-2
2. Courcelle, B., Olariu, S.: Upper bounds to the clique width of graphs. Discrete Appl. Math. **101**(1–3), 77–114 (2000). https://doi.org/10.1016/S0166-218X(99)00184-5
3. Damiand, G., Habib, M., Paul, C.: A simple paradigm for graph recognition: application to cographs and distance hereditary graphs. Theoret. Comput. Sci. **263**(1–2), 99–111 (2001). https://doi.org/10.1016/S0304-3975(00)00234-6. Combinatorics and computer science (Palaiseau, 1997)
4. Golumbic, M.C., Rotics, U.: On the clique-width of some perfect graph classes. Int. J. Found. Comput. Sci. **11**(3), 423–443 (2000). https://doi.org/10.1142/S0129054100000260. Selected papers from the Workshop on Theoretical Aspects of Computer Science (WG 99), Part 1 (Ascona)
5. Hammer, P.L., Maffray, F.: Completely separable graphs. Discrete Appl. Math. **27**(1–2), 85–99 (1990). Computational algorithms, operations research and computer science (Burnaby, BC, 1987)
6. Howorka, E.: A characterization of distance-hereditary graphs. Quart. J. Math. Oxford Ser. (2) **28**(112), 417–420 (1977)

Index

© The Author(s), under exclusive license to Springer Nature Switzerland AG 2022
C. Hoppen et al., *Locating Eigenvalues in Graphs*, SpringerBriefs in Mathematics,
https://doi.org/10.1007/978-3-031-11698-8

Printed in the United States
by Baker & Taylor Publisher Services